I0487742

Air Quality Modeling Technical Support Document: Tier 3 Motor Vehicle Emission and Standards

Air Quality Assessment Division
Office of Air Quality Planning and Standards
U.S. Environmental Protection Agency
Research Triangle Park, NC

United States
Environmental Protection
Agency

EPA-454/R-14-002
February 2014

Table of Contents

I.   Introduction...1

II.   Air Quality Modeling Platform...2
        A. Air Quality Model..2
        B. Model Domains and Grid Resolution...3
        C. Modeling Simulation Periods...4
        D. Modeling Scenarios..5
        E. Meteorological Input Data..7
        F. Initial and Boundary Conditions..9
        G. CMAQ Base Case Model Performance Evaluation..10

III.   CMAQ Model Results..10
        A. Impacts of Tier 3 Standards on Future 8-Hour Ozone Levels............................10
        B. Impacts of Tier 3 Standards on Future Annual $PM_{2.5}$ Levels........................12
        C. Impacts of Tier 3 Standards on Future 24-hour $PM_{2.5}$ Levels.......................15
        D. Impacts of Tier 3 Standards on Future Nitrogen Dioxide Levels......................17
        E. Impacts of Tier 3 Standards on Future Ambient Air Toxic Concentrations.......18
                1. Acetaldehyde...19
                2. Formaldehyde..20
                3. Benzene...21
                4. 1,3-Butadiene..22
                5. Acrolein...23
                6. Ethanol..25
                7. Naphthalene...26
        F. Air Toxics Population Metrics..27
        G. Impacts of Tier 3 Standards on Future Annual Nitrogen and Sulfur Deposition
             Levels...27
        H. Impacts of Tier 3 Standards on Future Visibility Levels..................................29

Appendices

i

List of Appendices

Appendix A.
Model Performance Evaluation for the 2007-Based Air Quality Modeling Platform

Appendix B.
8-Hour Ozone Design Values for Air Quality Modeling Scenarios

Appendix C.
Annual $PM_{2.5}$ Design Values for Air Quality Modeling Scenarios

Appendix D.
24-Hour $PM_{2.5}$ Design Values for Air Quality Modeling Scenarios

I. Introduction

This document describes the air quality modeling performed by EPA in support of the Tier 3 motor vehicle emission and fuel standards. A national scale air quality modeling analysis was performed to estimate the impact of the Tier 3 standards on future year annual and 24-hour $PM_{2.5}$ concentrations, daily maximum 8-hour ozone concentrations, annual nitrogen dioxide concentrations, annual nitrogen and sulfur deposition levels, annual ethanol and select annual and seasonal air toxic concentrations (formaldehyde, acetaldehyde, benzene, 1,3-butadiene, acrolein and naphthalene) as well as visibility impairment. To model the air quality benefits of this rule we used the Community Multiscale Air Quality (CMAQ) model.[1] CMAQ simulates the numerous physical and chemical processes involved in the formation, transport, and destruction of ozone, particulate matter and air toxics. In addition to the CMAQ model, the modeling platform includes the emissions, meteorology, and initial and boundary condition data which are inputs to this model.

Emissions and air quality modeling decisions are made early in the analytical process to allow for sufficient time required to conduct emissions and air quality modeling. For this reason, it is important to note that the inventories used in the air quality modeling and the benefits modeling, which are presented in Section 7.1 of the RIA, are slightly different than the final fuel and vehicle standard inventories presented in Section 7.2 of the RIA. However, the air quality inventories and the final rule inventories are generally consistent, so the air quality modeling adequately reflects the effects of the rule.

Air quality modeling was performed for five emissions cases: a 2007 base year, a 2018 reference case projection without the Tier 3 rule standards and a 2018 control case projection with Tier 3 standards in place, as well as a 2030 reference case projection without the Tier 3 rule standards and a 2018 control case projection with Tier 3 standards in place. The year 2007 was selected for the Tier 3 base year because this is the most recent year for which EPA had a complete national emissions inventory at the time of emission and air quality modeling.

The remaining sections of the Air Quality Modeling TSD are as follows. Section II describes the air quality modeling platform and the evaluation of model predictions of $PM_{2.5}$ and ozone using corresponding ambient measurements. In Section III we present the results of modeling performed for 2018 and 2030 to assess the impacts on air quality of the fuel and vehicle standards. Information on the development of emissions inventories for the Tier 3 Rule and the steps and data used in creating emissions inputs for air quality modeling can be found in the Emissions Inventory for Air Quality Modeling TSD (EITSD; EPA-HQ-OAR-2011-0135; EPA-454/R-14-003). The docket for this rulemaking also contains state/sector/pollutant emissions summaries for each of the emissions scenarios modeled.

---

[1] Byun, D.W., and K. L. Schere, 2006: Review of the Governing Equations, Computational Algorithms, and Other Components of the Models-3 Community Multiscale Air Quality (CMAQ) Modeling System. Applied Mechanics Reviews, Volume 59, Number 2 (March 2006), pp. 51-77.

II. Air Quality Modeling Platform

The 2007-based CMAQ modeling platform was used as the basis for the air quality modeling of the Tier 3 rule. This platform represents a structured system of connected modeling-related tools and data that provide a consistent and transparent basis for assessing the air quality response to projected changes in emissions. The base year of data used to construct this platform includes emissions and meteorology for 2007. The platform was developed by the U.S. EPA's Office of Air Quality Planning and Standards in collaboration with the Office of Research and Development and is intended to support a variety of regulatory and research model applications and analyses. This modeling platform and analysis is fully described below.

A. Air Quality Model

CMAQ is a non-proprietary computer model that simulates the formation and fate of photochemical oxidants, primary and secondary PM concentrations, acid deposition, and air toxics, over regional and urban spatial scales for given input sets of meteorological conditions and emissions. The CMAQ model version 5.0 was most recently peer-reviewed in September of 2011 for the U.S. EPA.[2] The CMAQ model is a well-known and well-respected tool and has been used in numerous national and international applications.[3,4,5] CMAQ includes numerous science modules that simulate the emission, production, decay, deposition and transport of organic and inorganic gas-phase and particle-phase pollutants in the atmosphere. This 2007 multi-pollutant modeling platform used the most recent multi-pollutant CMAQ code available at the time of air quality modeling (CMAQ version 5.0.1; multipollutant version[6]). CMAQ v5.0.1 reflects updates to version 4.7 to improve the underlying science which are detailed at http://www.cmascenter.org.[7,8]

---

[2] Brown, N., Allen, D., Amar, P., Kallos, G., McNider, R., Russell,, A., Stockwell, W. (September 2011). Final Report: Fourth Peer Review of the CMAQ Model, NERL/ORD/EPA. U.S. EPA, Research Triangle Park, NC., http://www.epa.gov/asmdnerl/Reviews/2011_CMAQ_Review_FinalReport.pdf. It is available from the Community Modeling and Analysis System (CMAS) as well as previous peer-review reports at: http://www.cmascenter.org.

[3] Hogrefe, C., Biswas, J., Lynn, B., Civerolo, K., Ku, J.Y., Rosenthal, J., et al. (2004). Simulating regional-scale ozone climatology over the eastern United States: model evaluation results. *Atmospheric Environment, 38(17)*, 2627-2638.

[4] United States Environmental Protection Agency. (2008). Technical support document for the final locomotive/marine rule: Air quality modeling analyses. Research Triangle Park, N.C.: U.S. Environmental Protection Agency, Office of Air Quality Planning and Standards, Air Quality Assessment Division.

[5] Lin, M., Oki, T., Holloway, T., Streets, D.G., Bengtsson, M., Kanae, S., (2008). Long range transport of acidifying substances in East Asia Part I: Model evaluation and sensitivity studies. Atmospheric Environment, 42(24), 5939-5955.

[6] CMAQ version 5.0.1 was released on July 2012. It is available from the Community Modeling and Analysis System (CMAS) website: http://www.cmascenter.org.

[7] Community Modeling and Analysis System (CMAS) website: http://www.cmascenter.org., RELEASE_NOTES for CMAQv5.0 - February 2012.

[8] Community Modeling and Analysis System (CMAS) website: http://www.cmascenter.org., RELEASE_NOTES for CMAQv5.0.1 - July 2012.

B. Model Domains and Grid Resolution

The CMAQ modeling analyses were performed for a domain covering the continental United States, as shown in Figure II-1. This domain has a parent horizontal grid of 36 kilometer (km) with a finer-scale 12 km grid. The model extends vertically from the surface to 50 millibars (approximately 17,600 meters) using a sigma-pressure coordinate system with 25 vertical layers. Air quality conditions at the outer boundary of the 36 km domain were taken from a global model and did not change over the simulations. In turn, the 36 km grid was only used to establish the incoming air quality concentrations along the boundaries of the 12 km grid. Only the finer grid data were used in determining the impacts of the Tier 3 standards. Table II-1 provides some basic geographic information regarding the CMAQ domains.

In addition to the CMAQ model, the Tier 3 modeling platform includes (1) emissions for the 2007 base year, 2018 reference and control case projection, 2030 reference and control case projection, (2) meteorology for the year 2007, and (3) estimates of intercontinental transport (i.e., boundary concentrations) from a global photochemical model. Using these input data, CMAQ was run to generate hourly predictions of ozone, $PM_{2.5}$ component species, nitrogen and sulfate deposition, nitrogen dioxide, ethanol and a subset of air toxics (formaldehyde, acetaldehyde, acrolein, benzene, 1,3-butadiene, and naphthalene) concentrations for each grid cell in the modeling domains. The development of 2007 meteorological inputs and initial and boundary concentrations are described below. The emissions inventories used in the Tier 3 air quality modeling are described in the EITSD found in the docket for this rule (EPA-HQ-OAR-2011-0135).

Table II-1. Geographic elements of domains used in Tier 3 modeling.

	CMAQ Modeling Configuration	
Grid Resolution	**36 km National Grid**	**12 km National Grid**
Map Projection	Lambert Conformal Projection	
Coordinate Center	97 deg W, 40 deg N	
True Latitudes	33 deg N and 45 deg N	
Dimensions	148 x 112 x 14	396 x 246 x 25
Vertical extent	25 Layers: Surface to 50 millibar level (see Table II-3)	

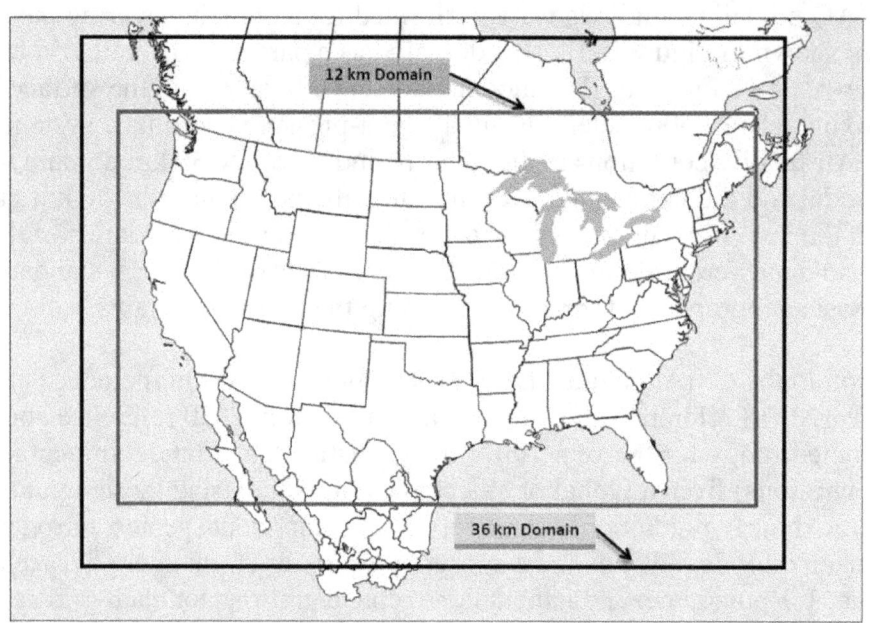

Figure II-1. Map of the CMAQ modeling domain. The black outer box denotes the 36 km national modeling domain; the purple inner box is the 12 km national fine grid modeling domain.

C. Modeling Simulation Periods

The 36 km and 12 km CMAQ modeling domains were modeled for the entire year of 2007. These annual simulations were performed in two half-year segments (i.e., January through June, July through December) for each emissions scenario. With this approach to segmenting an annual simulation we were able to reduce the overall throughput time for an annual simulation. The 36 km domain simulations included a "ramp-up" period, comprised of 10 days before the beginning of each half-year segment, to mitigate the effects of initial concentrations. For the 12 km domain simulations we used a 3-day ramp-up period for each half-year segment. The ramp-up periods are not considered as part of the output analyses. Fewer ramp-up days were used for the 12 km simulations because the initial concentrations were derived from the parent 36 km simulations.

For the 8-hour ozone results, we are only using modeling results from the period between May 1 and September 30, 2007. This 153-day period generally conforms to the ozone season across most parts of the U.S. and contains the majority of days with observed high ozone concentrations in 2007. Data from the entire year were utilized when looking at the estimation of $PM_{2.5}$, total nitrogen and sulfate deposition, nitrogen dioxide, ethanol, toxics and visibility impacts from this rulemaking.

D. Modeling Scenarios

As part of our analysis for this rulemaking, the CMAQ modeling system was used to calculate daily and annual $PM_{2.5}$ concentrations, 8-hour ozone concentrations, annual NO_2 concentrations, annual and seasonal air toxics concentrations, annual total nitrogen and sulfur deposition levels and visibility impairment for each of the following emissions scenarios:

2007 base year

2018 reference case projection without the Tier 3 fuel and vehicle standards

2018 control case projection with the Tier 3 fuel and vehicle standards

2030 reference case projection without the Tier 3 fuel and vehicle standards

2030 control case projection with the Tier 3 fuel and vehicle standards

Model predictions are used in a relative sense to estimate scenario-specific, future-year design values of $PM_{2.5}$ and ozone. For example, we compare a 2030 reference scenario (a scenario without the vehicle standards) to a 2030 control scenario which includes the vehicle standards. This is done by calculating the simulated air quality ratios between the 2030 future year simulation and the 2007 base. These predicted change ratios are then applied to ambient base year design values. The ambient air quality observations are average conditions, on a site-by-site basis, for a period centered around the model base year (i.e., 2005-2009). The raw model outputs are also used in a relative sense as inputs to the health and welfare impact functions of the benefits analysis. The difference between the 2030 reference case and 2030 control case was used to quantify the air quality benefits of the rule. Additionally, the differences in projected annual average $PM_{2.5}$ and seasonal average ozone were used to calculate monetized benefits by the BenMAP model (see Section 8.1.2 of the RIA).

The design value projection methodology used here followed EPA guidance[9] for such analyses. For each monitoring site, all valid design values (up to 3) from the 2005-2009 period were averaged together. Since 2007 is included in all three design value periods, this has the effect of creating a 5-year weighted average, where the middle year is weighted 3 times, the 2nd and 4th years are weighted twice, and the 1st and 5th years are weighted once. We refer to this as the 5-year weighted average value. The 5-year weighted average values were then projected to the future years that were analyzed for the proposed rule.

Concentrations of $PM_{2.5}$ in 2018 and 2030 were estimated by applying the modeled 2007-to-2018 and the modeled 2007-to-2030 relative change in $PM_{2.5}$ species to the 5 year weighted average (2005-2009) design values. Monitoring sites were included in the analysis if they had at least one complete design value in the 2005-2009 period. EPA followed the procedures recommended in the modeling guidance for projecting $PM_{2.5}$ by projecting individual $PM_{2.5}$

---

[9] U.S. EPA, 2007: Guidance on the Use of Models and Other Analyses for Demonstrating Attainment for Ozone, $PM_{2.5}$, and Regional Haze, Office of Air Quality Planning and Standards, Research Triangle Park, NC (EPA -454/B-07-002).

component species and then summing these to calculate the concentration of total $PM_{2.5}$. The $PM_{2.5}$ species are defined as sulfates, nitrates, ammonium, organic carbon mass, elemental carbon, crustal mass, water, and blank mass (a fixed value of 0.5 $\mu g/m^3$). EPA's Modeled Attainment Test Software (MATS) was used to calculate the future year design values. The software (including documentation) is available at: http://www.epa.gov/scram001/modelingapps_mats.htm.

To calculate 24-hour $PM_{2.5}$ design values, the measured 98th percentile concentrations from the 2005-2009 period at each monitor are projected to the future. The procedures for calculating the future year 24-hour $PM_{2.5}$ design values have been updated. The updates are intended to make the projection methodology more consistent with the procedures for calculating ambient design values.

A basic assumption of the old projection methodology is that the distribution of high measured days in the base period will be the same in the future. In other words, EPA assumed that the 98th-percentile day could only be displaced "from below" in the instance that a different day's future concentration exceeded the original 98th-percentile day's future concentration. This sometimes resulted in overstatement of future-year design values for 24-hour $PM_{2.5}$ at receptors whose seasonal distribution of highest-concentration 24-hour $PM_{2.5}$ days changed between the 2005-2009 period and the future year modeling.

In the revised methodology, we do not assume that the seasonal distribution of high days in the base period years and future years will remain the same. We project a larger set of ambient days from the base period to the future and then re-rank the entire set of days to find the new future 98th percentile value (for each year). More specifically, we project the highest 8 days per quarter (32 days per year) to the future and then re-rank the 32 days to derive the future year 98th percentile concentrations. More details on the methodology can be found in a guidance memo titled "Update to the 24 Hour $PM_{2.5}$ NAAQS Modeled Attainment Test" which can be found here: http://www.epa.gov/ttn/scram/guidance/guide/Update_to_the_24-hour_PM25_Modeled_Attainment_Test.pdf.

The future year 8-hour average ozone design values were calculated in a similar manner as the $PM_{2.5}$ design values. The May-to-September daily maximum 8-hour average concentrations from the 2007 base case and the 2018 and 2030 cases were used to project ambient design values to 2018 and 2030 respectively. The calculations used the base period 2005-2009 ambient ozone design value data for projecting future year design values. Relative response factors (RRF) for each monitoring site were calculated as the percent change in ozone on days with modeled ozone greater than 85 ppb[10].

We also conducted an analysis to compare the absolute and percent differences between the 2018 control case and the 2018 reference case as well as the 2030 control case and the 2030 reference case for annual and seasonal nitrogen dioxide, ethanol, formaldehyde, acetaldehyde, benzene, 1,3-butadiene, acrolein, and naphthalene as well as annual nitrate and sulfate

---

[10] If there are less than 5 days > 70 ppb for a site, then the threshold is lowered in 1 ppb increments to as low as 60 ppb. If there are not 5 days > 60 ppb, then the site is excluded. If a county has no sites that meet the 70 ppb threshold, then the county design value is calculated from the sites that meet the 60 ppb threshold.

deposition. These data were not compared in a relative sense due to the limited observational data available.

E. Meteorological Input Data

The gridded meteorological input data for the entire year of 2007 were derived from simulations of the Weather Research and Forecasting Model (WRF) version 3.3, Advanced Research WRF (ARW) core[11] for the entire year of 2007 over model domains that are slightly larger than those shown in Figure II-1. Meteorological model input fields were prepared separately for the 36 km and 12 km domains shown in Figure II-1. The WRF simulations were run on the same map projection as CMAQ.

The 36 km and 12 km meteorological model runs configured similarly. The selections for key WRF physics options are shown below[12]:

- Pleim-Xiu PBL and land surface schemes
- Asymmetric Convective Model version 2 planetary boundary layer scheme
- Kain-Fritsh cumulus parameterization
- Morrison double moment microphysics
- RRTMG longwave and shortwave radiation schemes

Three dimensional analysis nudging for temperature, wind, and moisture was applied above the boundary layer only. The meteorological simulations were conducted in 5.5 day blocks with soil moisture and temperature carried from one block to the next via the ipxwrf program.[13] Landuse and land cover data are based on the U.S. Geological Survey (USGS) data. The 36km and 12km meteorological modeling domains contained 35 vertical layers with an approximately 19 m deep surface layer and a 50 millibar top. The WRF and CMAQ vertical structures are shown in Table II-3 and do not vary by horizontal grid resolution.

Table II-3. Vertical layer structure for WRF and CMAQ (heights are layer top).

CMAQ Layers	WRF Layers	Sigma P	Approximate Height (m)
25	35	0.0000	17,556
	34	0.0500	14,780
24	33	0.1000	12,822
	32	0.1500	11,282
23	31	0.2000	10,002

[11] Skamarock, W.C., Klemp, J.B., Dudhia, J., Gill, D.O., Barker, D.M., Duda, M.G., Huang, X., Wang, W., Powers, J.G., 2008. A Description of the Advanced Research WRF Version 3.

[12] Gilliam, R.C., Pleim, J.E., 2010. Performance Assessment of New Land Surface and Planetary Boundary Layer Physics in the WRF-ARW. Journal of Applied Meteorology and Climatology 49, 760-774.

[13] Gilliam, R.C., Pleim, J.E., 2010. Performance Assessment of New Land Surface and Planetary Boundary Layer Physics in the WRF-ARW. Journal of Applied Meteorology and Climatology 49, 760-774.

	30	0.2500	8,901
22	29	0.3000	7,932
	28	0.3500	7,064
21	27	0.4000	6,275
	26	0.4500	5,553
20	25	0.5000	4,885
	24	0.5500	4,264
19	23	0.6000	3,683
18	22	0.6500	3,136
17	21	0.7000	2,619
16	20	0.7400	2,226
15	19	0.7700	1,941
14	18	0.8000	1,665
13	17	0.8200	1,485
12	16	0.8400	1,308
11	15	0.8600	1,134
10	14	0.8800	964
9	13	0.9000	797
	12	0.9100	714
8	11	0.9200	632
	10	0.9300	551
7	9	0.9400	470
	8	0.9500	390
6	7	0.9600	311
5	6	0.9700	232
4	5	0.9800	154
	4	0.9850	115
3	3	0.9900	77
2	2	0.9950	38
1	1	0.9975	19
0	0	1.0000	0

The 2007 meteorological outputs from the 36km and 12km WRF sets were processed to create model-ready inputs for CMAQ using the Meteorology-Chemistry Interface Processor (MCIP), version 4.1.2.[14,15]

---

[14] Byun, D.W., and Ching, J.K.S., Eds, 1999. Science algorithms of EPA Models-3 Community Multiscale Air Quality (CMAQ modeling system, EPA/600/R-99/030, Office of Research and Development).
[15] Otte, T.L., Pleim, J.E., 2010. The Meteorology-Chemistry Interface Processor (MCIP) for the CMAQ modeling system: updates through MCIPv3.4.1. Geoscientific Model Development 3, 243-256.

Before initiating the air quality simulations, it is important to identify the biases and errors associated with the meteorological modeling inputs. The 2007 WRF model performance evaluations used an approach which included a combination of qualitative and quantitative analyses to assess the adequacy of the WRF simulated fields. The qualitative aspects involved comparisons of the model-estimated synoptic patterns against observed patterns from historical weather chart archives. Additionally, the evaluations compared spatial patterns of monthly average rainfall and monthly maximum planetary boundary layer (PBL) heights. The operational evaluation included statistical comparisons of model/observed pairs (e.g., mean bias, mean (gross) error, fractional bias, and fractional error[16]) for multiple meteorological parameters. For this portion of the evaluation, five meteorological parameters were investigated: temperature, humidity, shortwave downward radiation, wind speed, and wind direction. The 36 km and 12 km WRF evaluations are described elsewhere.[17] The results of these analyses indicate that the bias and error values associated with all three sets of 2007 meteorological data were generally within the range of past meteorological modeling results that have been used for air quality applications.

F. Initial and Boundary Conditions

The lateral boundary concentrations are provided by a three-dimensional global atmospheric chemistry model, the GEOS-CHEM[18,19] model (standard version 8-03-02 with version 8-02-03 chemistry). The global GEOS-CHEM model simulates atmospheric chemical and physical processes driven by assimilated meteorological observations from the NASA's Goddard Earth Observing System (GEOS-5). This model was run for 2007 with a grid resolution of 2.0 degree x 2.5 degree (latitude-longitude) and 46 vertical layers up to 0.01 hPa. The predictions were processed using the GEOS-2-CMAQ tool[20,21] and used to provide one-way dynamic boundary conditions at one-hour intervals. The ozone from these GEOS-Chem runs was evaluated by comparing to satellite vertical profiles and ground-based measurements and found acceptable model performance.

Initial conditions were extracted from a slightly older model simulation using GEOS-CHEM version 8-02-03. The model simulation from which the initial conditions were extracted was also run with a grid resolution of 2.0 of 2.0 degree x 2.5 degree (latitude-longitude) and 46 vertical layers. A GEOS-Chem evaluation was conducted for the purpose of validating the 2007 GEOS-Chem simulation outputs for their use as inputs to the CMAQ modeling system. This

[16] Boylan, J.W., Russell, A.G., 2006. PM and light extinction model performance metrics, goals, and criteria for three-dimensional air quality models. Atmospheric Environment 40, 4946-4959.

[17] Misenis, Chris Meteorological Model Performance Evaluation for the Annual 2007 Simulation WRF v3.3, USEPA/OAQPS, July 15, 2012.

[18] Yantosca, B.,2004. GEOS-CHEMv7-01-02 User's Guide, Atmospheric Chemistry Modeling Group, Harvard University, Cambridge, MA, October 15, 2004.

[19] Le Sager, P. Yantosca, B., Carouge, C. (2008). GEOS-CHEM v8-01-02 User's Guide, Atmospheric Chemistry Modeling Group, Harvard University, Cambridge, MA, December 18, 2008.

[20] Akhtar, F., Henderson, B., Appel, W., Napelenok, S., Hutzell, B., Pye, H., Foley, K.,2012. Multiyear Boundary Conditions for CMAQ 5.0 from GEOS-Chem with Secondary Organic Aerosol Extensions, 11th annual Community Modeling and Analysis System conference, Chapel Hill, NC, October 2012.

[21] Henderson, B.H., Akhtar, F., Pye, H.O.T., Napelenok, S.L., Hutzell, W.T., 2013. A database and tool for boundary conditions for regional air quality modeling: description and evaluation, Geoscientific Model Development Discussions, 6, 4665-4704.

evaluation included reproducing GEOS-Chem evaluation plots reported in the literature for previous versions of the model.[22]

G. CMAQ Base Case Model Performance Evaluation

The CMAQ predictions for ozone, fine particulate matter, sulfate, nitrate, ammonium, organic carbon, elemental carbon, a selected subset of toxics, and nitrogen and sulfur deposition from the 2007 base year evaluation case were compared to measured concentrations in order to evaluate the performance of the modeling platform for replicating observed concentrations. This evaluation was comprised of statistical and graphical comparisons of paired modeled and observed data. Details on the model performance evaluation including a description of the methodology, the model performance statistics, and results are provided in Appendix A.

III. CMAQ Model Results

As described above, we performed a series of air quality modeling simulations for the continental U.S in order to assess the impacts of the Tier 3 standards. We looked at impacts on future ambient levels of $PM_{2.5}$, ozone and NO_2, as well as changes in ambient concentrations of ethanol and the following air toxics: acetaldehyde, acrolein, benzene, 1,3-butadiene, naphthalene and formaldehyde. The air quality modeling results also include impacts on deposition of nitrogen and sulfur and on visibility levels due to this rule. In this section, we present the air quality modeling results for the 2018 Tier 3 control case relative to the 2018 reference case as well as the 2030 Tier 3 control case relative to the 2030 reference case.

A. Impacts of Tier 3 Standards on Future 8-Hour Ozone Levels

This section summarizes the results of our modeling of ozone air quality impacts in the future with the Tier 3 fuel and vehicle standards. Specifically, for the years 2018 and 2030 we compare a reference scenario (a scenario without the proposed Tier 3 standards) to a control scenario which includes the Tier 3 standards. Our modeling indicates that there will be substantial decreases in ozone across most of the country as a result of the Tier 3 standards.

Figure III-1 and Figure III-2 present the changes in 8-hour ozone design value concentrations between the reference case and the control case in 2018 and 2030 respectively.[23] Appendix B details the state and county 8-hour maximum ozone design values for the ambient baseline and the 2018 and 2030 future reference and control cases.

---

[22] Lam, Y.F., Fu, J.S., Jacob, D.J., Jang, C., Dolwick, P., 2010 2006-2008 GEOS-Chem for CMAQ Initial and Boundary Conditions. 9th Annual CMAS Conference, October 11-13, 2010, Chapel Hill, NC.

[23] An 8-hour ozone design value is the concentration that determines whether a monitoring site meets the 8-hour ozone NAAQS. The full details involved in calculating an 8-hour ozone design value are given in Appendix I of 40 CFR part 50.

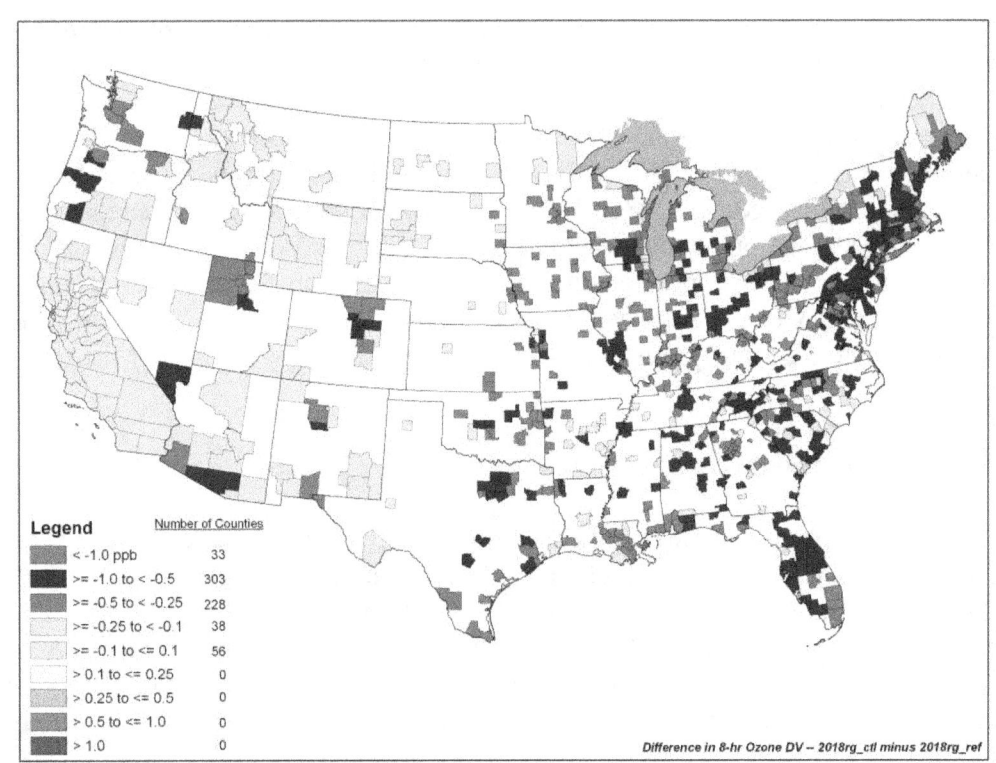

Figure III-1. Projected Change in 2018 8-hour Ozone Design Values Between the Reference Case and Control Case

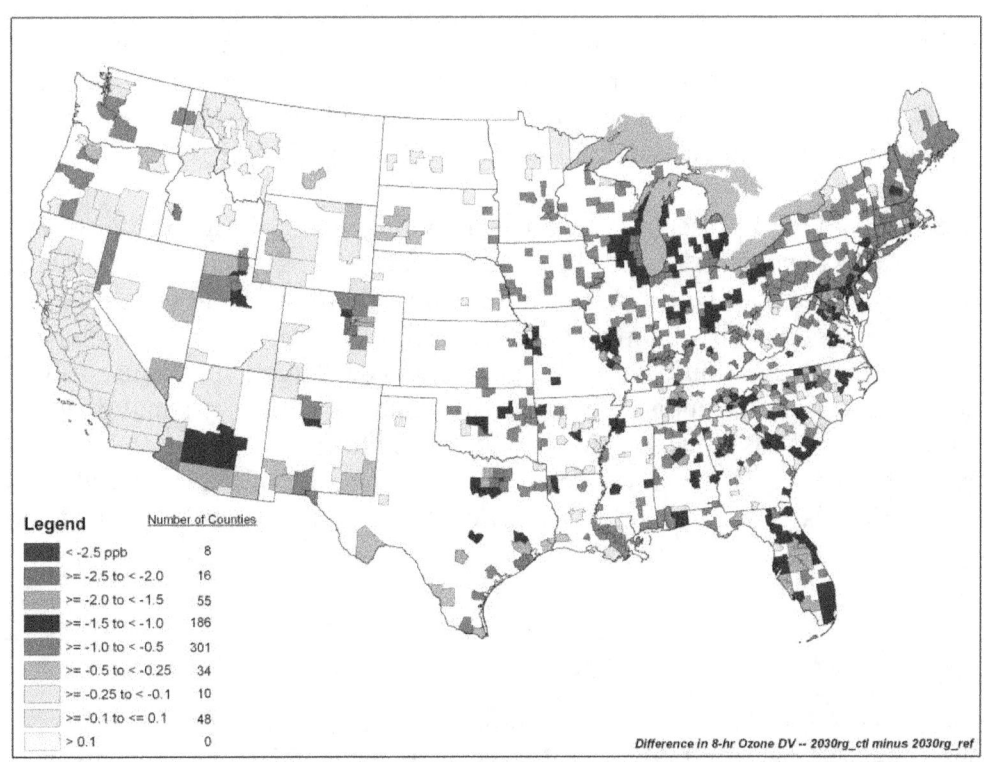

Figure III-2. Projected Change in 2030 8-hour Ozone Design Values Between the Reference Case and Control Case

As can be seen in Figure III-1, the majority of the design value decreases in 2018 are between 0.5 and 1.0 ppb. There are also 33 counties with projected 8-hour ozone design value decreases of more than 1 ppb; these counties are generally in urban areas in states that have not adopted California LEV III standards. The maximum projected decrease in an 8-hour ozone design value in 2018 is 1.56 ppb in Henry County, Georgia near Atlanta. Figure III-2 presents the ozone design value changes for 2030. In 2030 the ozone design value decreases are larger than in 2018; most decreases are projected to be between 0.5 and 1.0 ppb, but over 250 more counties have design values with projected decreases greater than 1.5 ppb. The maximum projected decrease in an 8-hour ozone design value in 2030 is 2.8 ppb in Gwinnett County, Georgia, the northeastern part of the Atlanta metropolitan area.

B. Impacts of Tier 3 Standards on Future Annual PM$_{2.5}$ Levels

This section summarizes the results of our modeling of annual average PM$_{2.5}$ air quality impacts in the future due to the Tier 3 fuel and vehicle standards. Specifically, for the years 2018 and 2030 we compare a reference scenario (a scenario without the standards) to a control scenario that includes the standards. Our modeling indicates that by 2030 annual PM$_{2.5}$ design values in the majority of the modeled counties would decrease due to the standards. The decreases in annual PM$_{2.5}$ design values are likely due to the projected reductions in primary PM$_{2.5}$, NO$_X$, SO$_X$ and VOC emissions (see Section 7.2.1 in the RIA). Note that the air quality modeling used inventories that included an increase in direct PM$_{2.5}$ emissions in the West and Pacific Northwest that is an artifact of a difference in fuel properties that isn't real.[24] Although in most areas this direct PM$_{2.5}$ increase is outweighed by reductions in secondary PM$_{2.5}$, the air quality modeling does predict ambient PM$_{2.5}$ increases in a few places in the West and Pacific Northwest. These modeled increases are a result of the inventory issue, and we do not expect them to actually occur. Appendix C details the state and county annual PM$_{2.5}$ design values for the ambient baseline and the 2018 and 2030 future reference and control cases.

Figure III-3 and III-4 presents the changes in annual PM$_{2.5}$ design values in 2018 and 2030 respectively.[25] As shown in Figure III-3, we project that in 2018 over 200 counties will have design value decreases of between 0.01 µg/m^3 and 0.05 µg/m^3. These counties tend to be in urban areas in states that have not adopted California LEV III standards. The maximum projected decrease in a 2018 annual PM$_{2.5}$ design value is 0.04 µg/m^3 in Waukesha County, Wisconsin and Cook County, Illinois. There are two counties with very small projected increases in their annual PM$_{2.5}$ design values in 2018: Lewis & Clark County, Montana, and Gallatin County, Montana. These projected increases are a result of the issue with the air quality modeling inventories discussed in Section 7.2.1.1 of the RIA, and we do not expect these increases will occur.

---

[24] The issue is with the way that some of the fuel property data, specifically E200/E300 and T50/T90, matched up in the fuel compliance database in the West and Pacific Northwest, see Section 7.2.1.1 for additional information.
[25] An annual PM$_{2.5}$ design value is the concentration that determines whether a monitoring site meets the annual NAAQS for PM$_{2.5}$. The full details involved in calculating an annual PM$_{2.5}$ design value are given in appendix N of 40 CFR part 50.

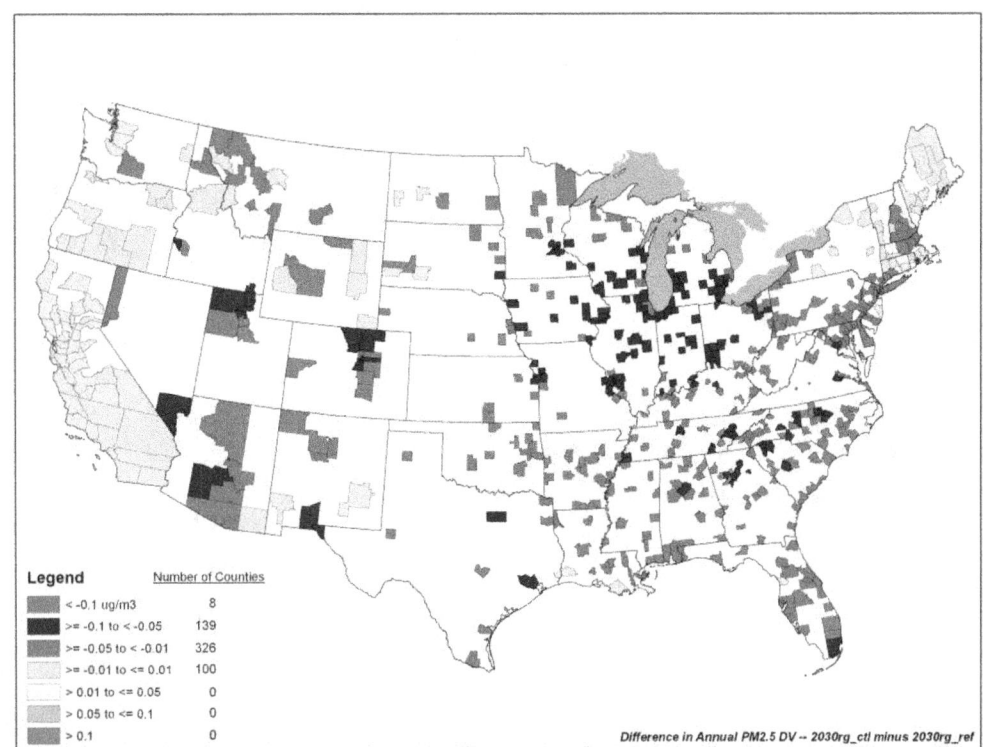

Legend	Number of Counties
< -0.1 ug/m3	8
>= -0.1 to < -0.05	139
>= -0.05 to < -0.01	326
>= -0.01 to <= 0.01	100
> 0.01 to <= 0.05	0
> 0.05 to <= 0.1	0
> 0.1	0

Difference in Annual PM2.5 DV -- 2030rg_ctl minus 2030rg_ref

Figure III-4 presents the annual $PM_{2.5}$ design value changes in 2030. The annual $PM_{2.5}$ design value decreases in 2030 are larger than the decreases in 2018; most design values are projected to decrease between 0.01 and 0.05 $\mu g/m^3$ and over 140 additional counties have projected design value decreases greater than 0.05 $\mu g/m^3$. The maximum projected decrease in an annual $PM_{2.5}$ design value in 2030 is 0.15 $\mu g/m^3$ in Milwaukee County, Wisconsin.

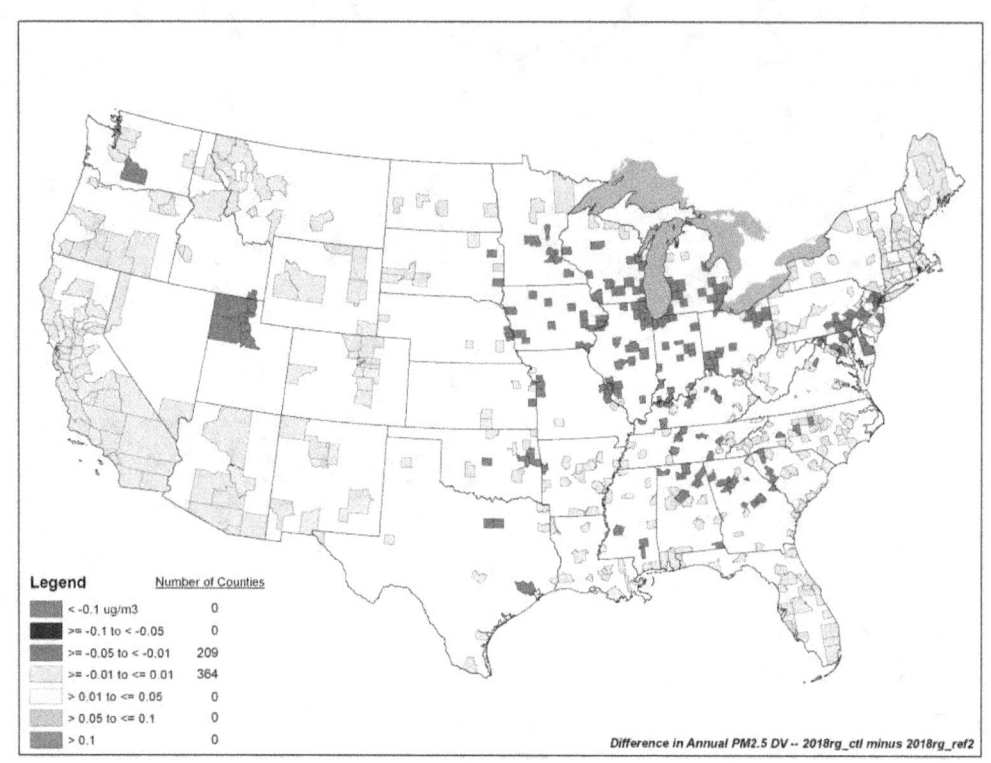

Figure III-3. Projected Change in 2018 Annual PM$_{2.5}$ Design Values Between the Reference Case and Control Case

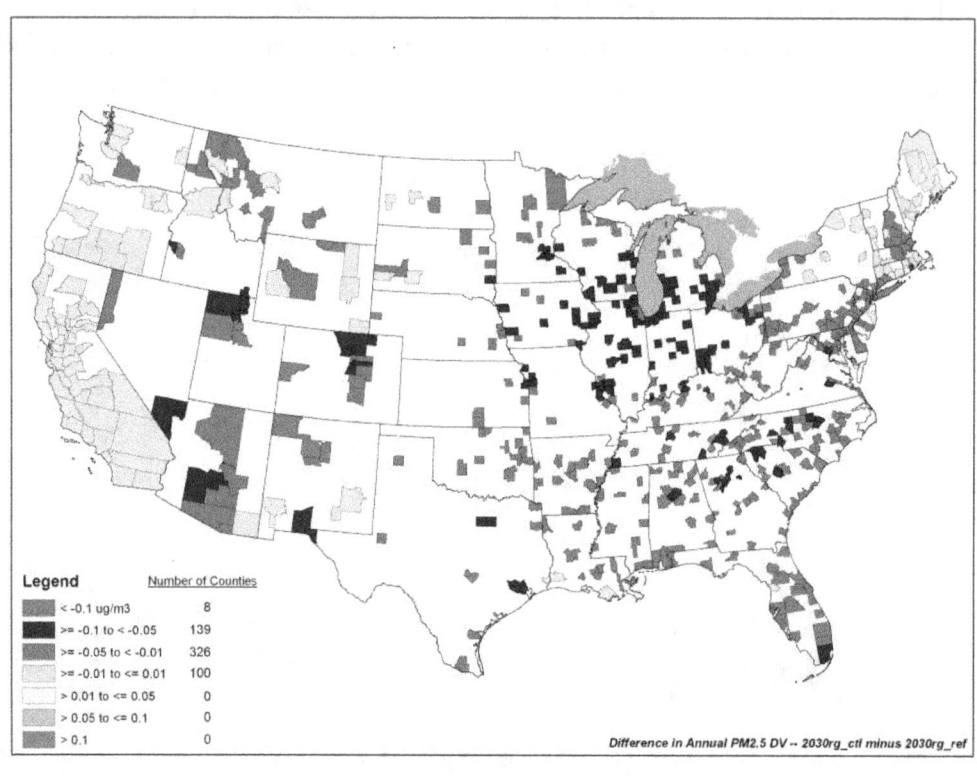

Figure III-4. Projected Change in 2030 Annual PM$_{2.5}$ Design Values Between the Reference Case and Control Case

14

C. Impacts of Tier 3 Standards on Future 24-hour PM$_{2.5}$ Levels

This section summarizes the results of our modeling of 24-hour PM$_{2.5}$ air quality impacts in the future due to the Tier 3 rule. Specifically, for the years 2018 and 2030 we compare a reference scenario (a scenario without the proposed standards) to a 2030 control scenario that includes the standards. Our modeling indicates that 24-hour PM$_{2.5}$ design values in the majority of the modeled counties would decrease due to the standards. The decreases in 24-hour PM$_{2.5}$ design values are likely due to the projected reductions in primary PM$_{2.5}$, NO$_X$, SO$_X$ and VOCs. As described in Section 7.2.1.1 of the RIA, the air quality modeling used inventories that include an increase in direct PM$_{2.5}$ emissions in the West and Pacific Northwest that is an artifact of a difference in fuel properties that isn't real.[26] Although in most areas this direct PM$_{2.5}$ increase is outweighed by reductions in secondary PM$_{2.5}$, the air quality modeling does predict ambient PM$_{2.5}$ increases in a few places in the West and Pacific Northwest. These modeled increases are a result of the inventory issue, and we do not expect them to actually occur. Ambient PM$_{2.5}$ projections are discussed in more detail below. Figures III-5 and III-6 present the changes in 24-hour PM$_{2.5}$ design values in 2018 and 2030 respectively.[27] Appendix D details the state and county 24-hour PM$_{2.5}$ design values for the ambient baseline and the future reference and control cases.

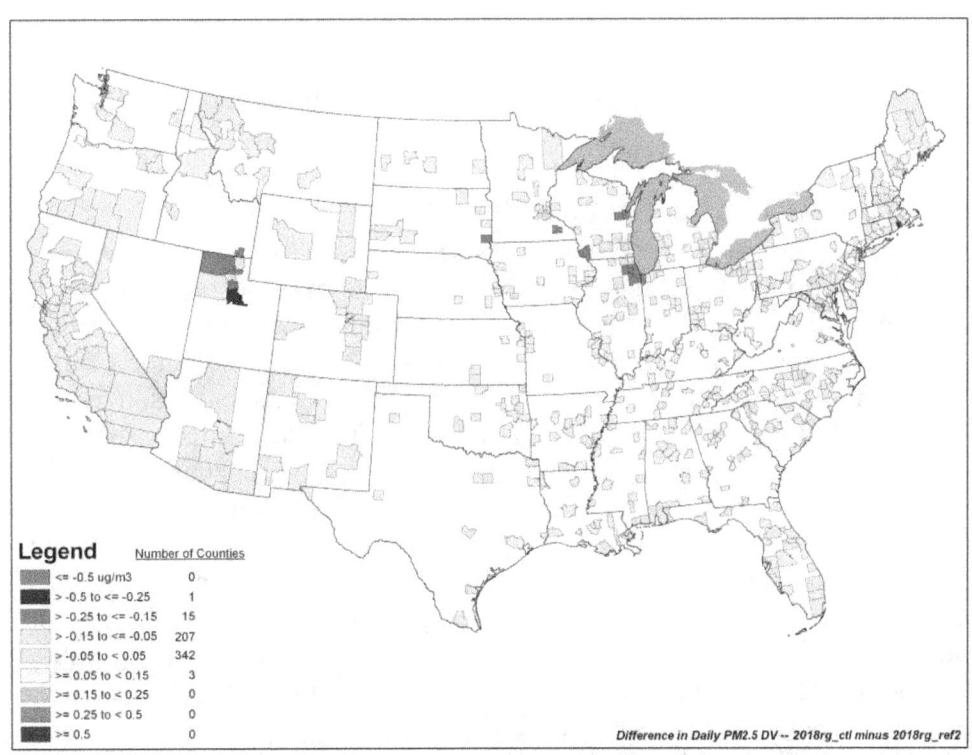

Figure III-5. Projected Change in 2018 24-hour PM$_{2.5}$ Design Values Between the Reference Case and the Control Case

---

[26] The issue is with the way that some of the fuel property data, specifically E200/E300 and T50/T90, matched up in the fuel compliance database in the West and Pacific Northwest, see Section 7.2.1.1 for additional information.
[27] A 24-hour PM$_{2.5}$ design value is the concentration that determines whether a monitoring site meets the 24-hour NAAQS for PM$_{2.5}$. The full details involved in calculating a 24-hour PM$_{2.5}$ design value are given in appendix N of 40 CFR part 50.

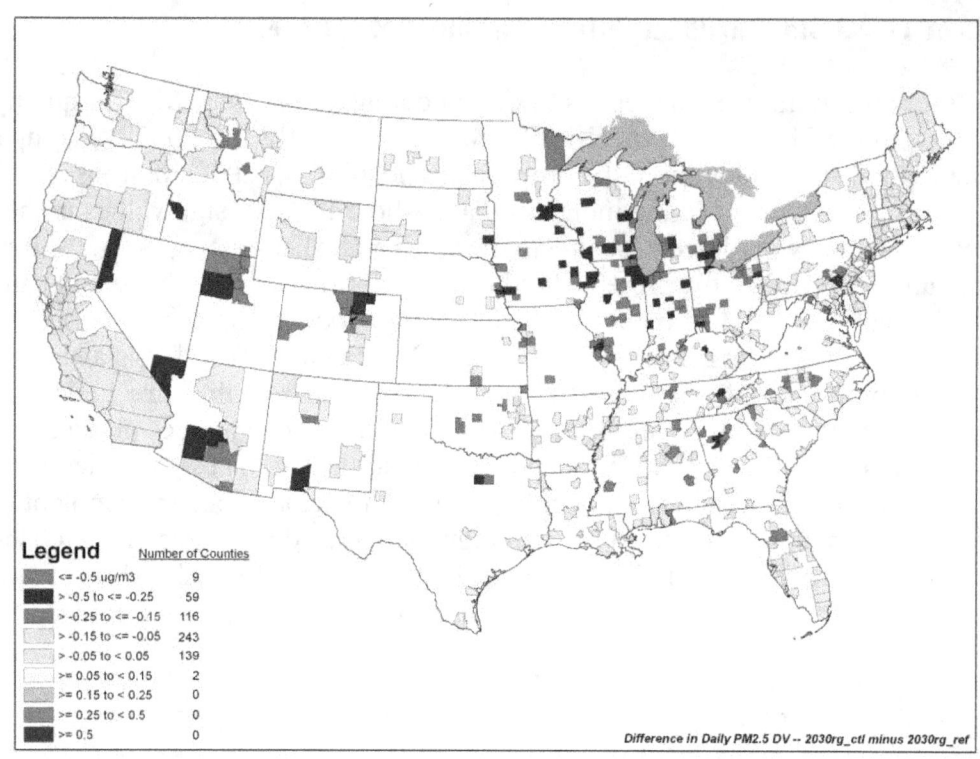

Figure III-6. Projected Change in 2030 24-hour PM$_{2.5}$ Design Values Between the Reference Case and the Control Case

As shown in Figure III-5, in 2018 there are 16 counties with projected 24-hour PM$_{2.5}$ design value decreases greater than 0.15 µg/m^3. These counties are in urban areas in states that have not adopted California LEV III standards. The maximum projected decrease in a 2018 24-hour PM$_{2.5}$ design value is 0.30 µg/m^3 in Utah County, Utah. There are three counties with projected increases in their 24-hour PM$_{2.5}$ design values in 2018: Washington County, Oregon; King County, Washington; and Sheridan County, Wyoming. These projected increases are a result of the issue with the air quality modeling emissions inventories discussed in Section 7.2.1.1 of the Tier 3 RIA, and we do not expect these increases will occur. Figure III-6 presents the 24-hour PM$_{2.5}$ design value changes in 2030. In 2030 the 24-hour PM$_{2.5}$ design value decreases are larger; most design values are projected to decrease between 0.05 and 0.15 µg/m^3 and over 50 counties have projected design value decreases greater than 0.25 µg/m^3. The maximum projected decrease in a 24-hour PM$_{2.5}$ design value in 2030 is 0.8 µg/m^3 in Salt Lake County, Utah. As shown in Figure III-6, design values in 9 counties are projected to decrease by more than 0.5 µg/m^3. These counties are in Utah, Idaho, Colorado and Wisconsin. There are two counties with projected increases in their 24-hour PM$_{2.5}$ design values in 2030: King County, Washington, and Pierce County, Washington. These projected increases are a result of the issue with the air quality modeling emissions inventories discussed in Section 7.2.1.1 of the RIA and we do not expect these increases will occur.

D. Impacts of Tier 3 Standards on Future Nitrogen Dioxide Levels

This section summarizes the results of our modeling of annual average nitrogen dioxide (NO_2) air quality impacts in the future due to the final Tier 3 standards. Specifically, for the years 2018 and 2030 we compare a reference scenario (a scenario without the Tier 3 standards) to a control scenario that includes the Tier 3 standards. Figure III-7 and Figure III-8 present the changes in annual NO_2 concentrations in 2018 and 2030 respectively.

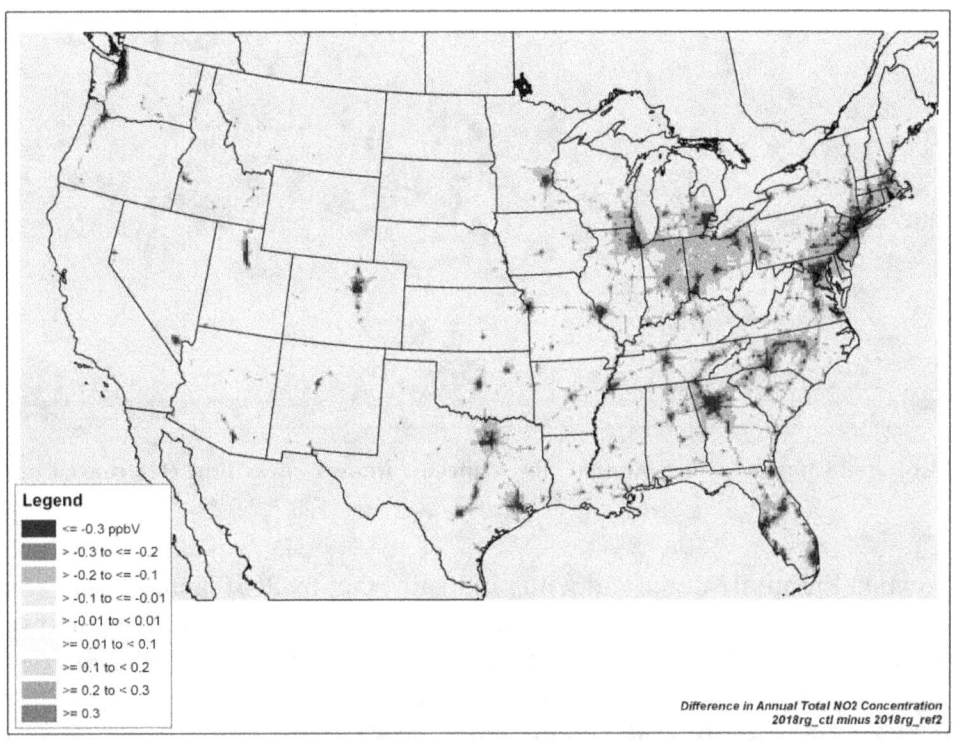

Figure III-7. Projected Change in 2018 Annual NO₂ Concentrations Between the Reference Case and Control Case

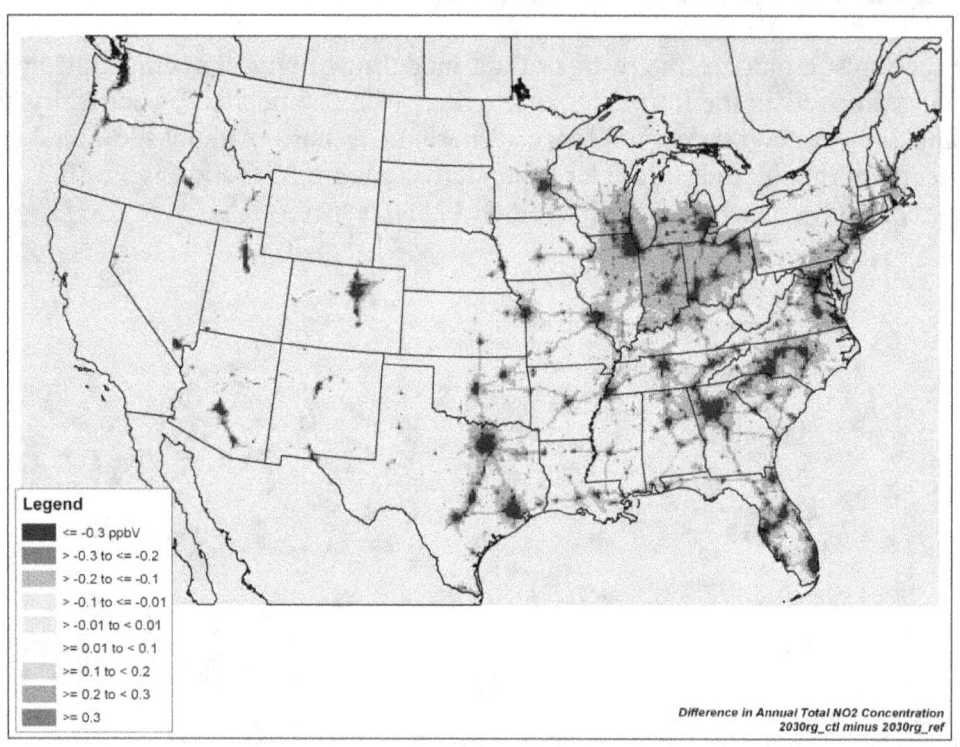

Legend

- <= -0.3 ppbV
- > -0.3 to <= -0.2
- > -0.2 to <= -0.1
- > -0.1 to <= -0.01
- > -0.01 to < 0.01
- >= 0.01 to < 0.1
- >= 0.1 to < 0.2
- >= 0.2 to < 0.3
- >= 0.3

Difference in Annual Total NO2 Concentration
2030rg_ctl minus 2030rg_ref

Figure III-8. Projected Change in 2030 Annual NO₂ Concentrations Between the Reference Case and Control Case

As shown in Figure III-8, our modeling indicates that by 2030 annual NO$_2$ concentrations in the majority of the country would decrease less than 0.1 ppb due to this rule. However, decreases in annual NO$_2$ concentrations are greater than 0.3 ppb in most urban areas. These emissions reductions would also likely decrease 1-hour NO$_2$ concentrations and help any potential nonattainment areas to attain and maintain the standard.

E. Impacts of Tier 3 Standards on Future Ambient Air Toxic Concentrations

The following sections summarize the results of our modeling of air toxics impacts in the future from the Tier 3 fuel and vehicle emission standards. We focus on air toxics which were identified as national and regional-scale cancer and noncancer risk drivers in the 2005 National-Scale Air Toxics Assessment (NATA)[28] and were also likely to be significantly impacted by the standards. These compounds include benzene, 1,3-butadiene, formaldehyde, acetaldehyde, naphthalene, and acrolein. Impacts on ethanol concentrations were also included in our analyses. Our modeling indicates that the impacts of the standards include generally small decreases in ambient concentrations of air toxics, with the greatest reductions in urban areas. Air toxics pollutants dominated by primary emissions (or a decay product of a directly emitted pollutant), such as benzene and 1,3-butadiene, have the largest impacts. Air toxics that primarily result

---

[28] U.S. EPA. (2011) 2005 National-Scale Air Toxics Assessment. http://www.epa.gov/ttn/atw/nata2005/. Docket EPA-HQ-OAR-2011-0135.

from photochemical transformation, such as formaldehyde and acetaldehyde, are not impacted as much as those dominated by direct emissions. Our modeling shows decreases in ambient air toxics concentrations for both 2018 and 2030. Reductions are greater in 2030, when Tier 3 cars and trucks would contribute nearly 90 percent of fleet-wide vehicle miles travelled, than in 2018. However, our 2018 modeling projects there would be small immediate reductions in ambient concentrations of air toxics due to the sulfur controls that take effect in 2017. Furthermore, the full reduction of the vehicle program would be realized after 2030, when the fleet has fully turned over to Tier 3 vehicles. Because overall impacts are relatively small in both future years, we concluded that assessing exposure to ambient concentrations and conducting a quantitative risk assessment of air toxic impacts was not warranted. However, we did develop population metrics, including the population living in areas with increases or decreases in concentrations of various magnitudes.

1. Acetaldehyde

Air quality modeling shows annual percent changes in ambient concentrations of acetaldehyde of generally less than 1 percent across the U.S., although the proposal may decrease acetaldehyde concentrations in some urban areas by 1 to 2.5 percent in 2030. Changes in ambient concentrations of acetaldehyde are generally in the range of 0.01 $\mu g/m^3$ to -0.01 $\mu g/m^3$ with decreases happening in the more populated areas and increases happening in more rural areas.

The complex photochemistry associated with NO_X emissions and acetaldehyde formation appears to be the explanation for the split between increased rural concentrations and decreased urban concentrations. In the atmosphere, acetaldehyde precursors react with NO_X to form peroxyacylnitrate (PAN). Reducing NO_X allows acetaldehyde precursors to be available to form acetaldehyde instead. This phenomenon is more prevalent in rural areas where NO_X is low. The chemistry involved is further described by a recent study done by EPA's Office of Research and Development and Region 3 evaluating the complex effects of reducing multiple emissions on reactive air toxics and criteria pollutants.[29]

---

[29] Luecken, D,J, Clmorel, A.J. 2008. Codependencies of Reactive Air Toxic and Criteria Pollutants on Emission Reductions. J. Air & Waste Manage. Assoc. 58:693–701. DOI:10.3155/1047-3289.58.5.693

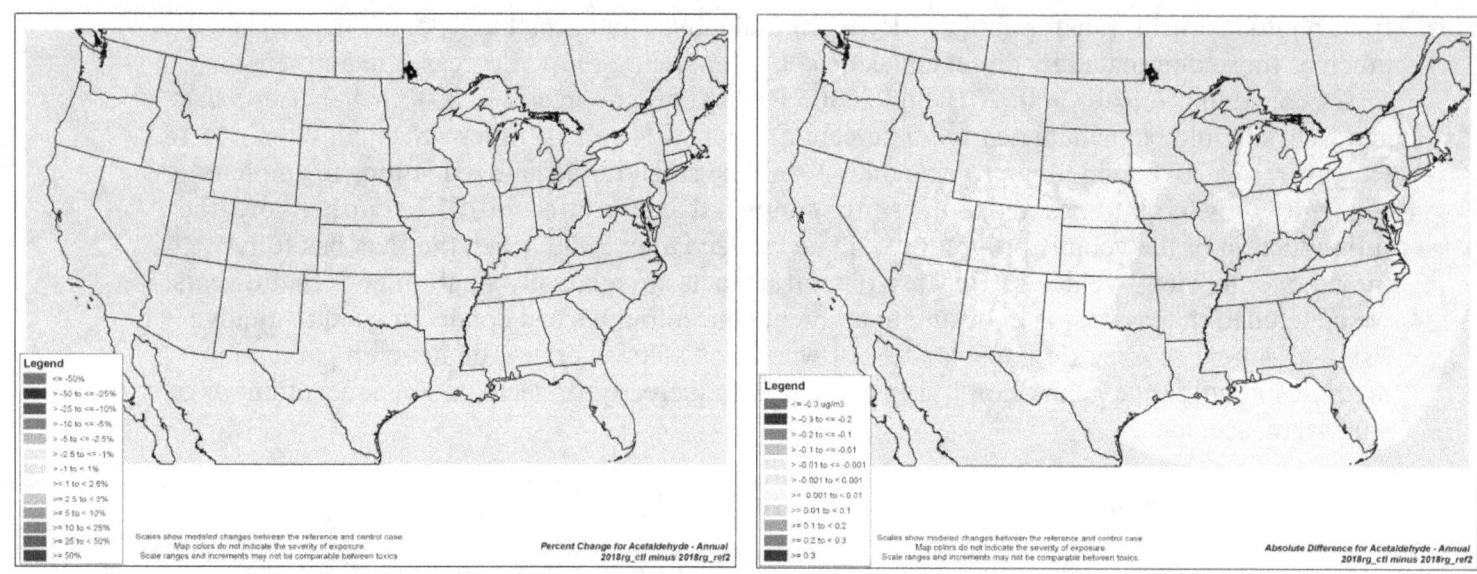

Figure III-9. Changes in Annual Acetaldehyde Ambient Concentrations Between the Reference Case and the Control Case in 2018: Percent Changes (left) and Absolute Changes in µg/m³ (right)

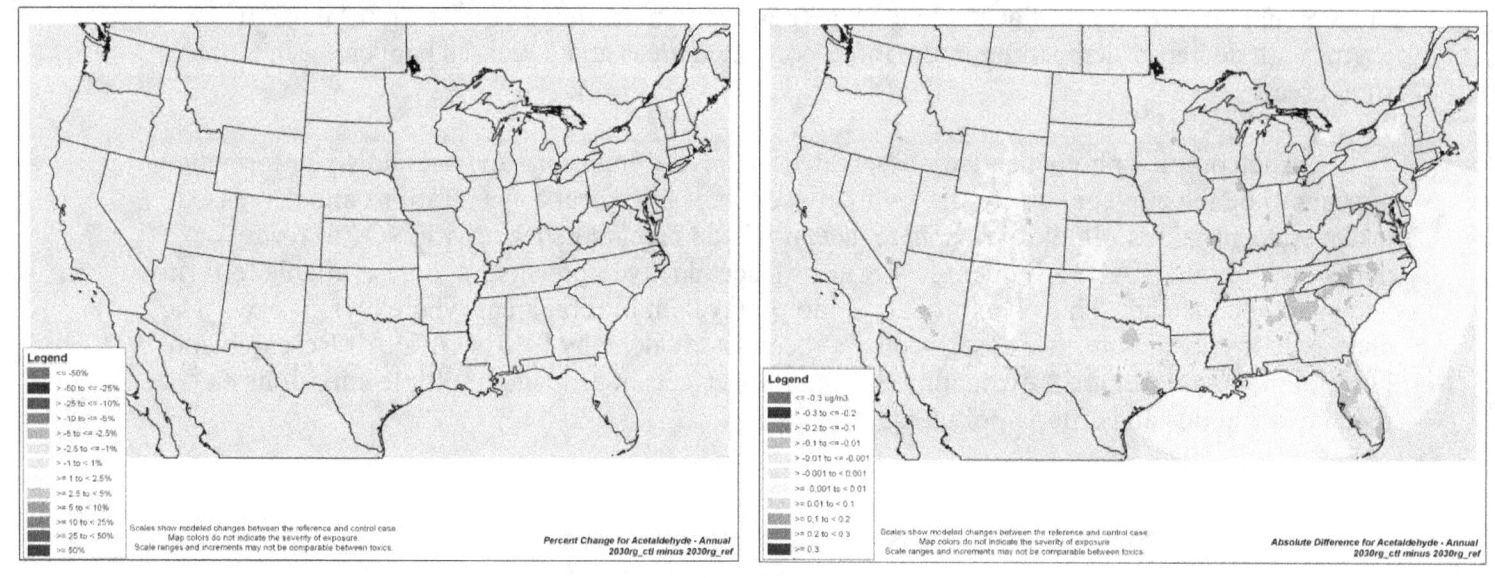

Figure III-10. Changes in Annual Acetaldehyde Ambient Concentrations Between the Reference Case and the Control Case in 2030: Percent Changes (left) and Absolute Changes in µg/m³ (right)

2. Formaldehyde

Our modeling projects that formaldehyde concentrations would slightly decrease in parts of the country (mainly urban areas) as a result of the Tier 3 final rule. As shown in Figure III-11.III-11 and Figure III-12, annual percent changes in ambient concentrations of formaldehyde are less than 1 percent across much of the country for 2018 but are on the order of 1 to 5 percent in 2030 in some urban areas as a result of the rule. Figure III-11.III-11 and Figure III-12 also show that absolute changes in ambient concentrations of formaldehyde are generally between 0.001 and 0.01 µg/m³ in both years, with some areas as high as 0.1 µg/m³ in 2030.

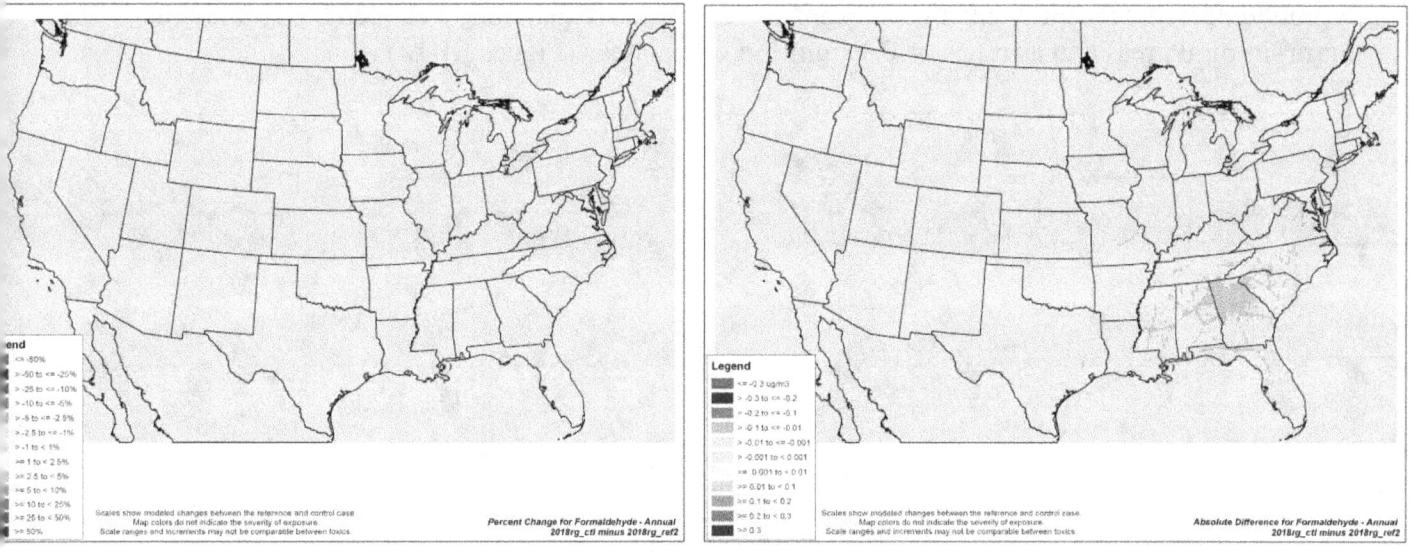

Figure III-11. Changes in Formaldehyde Ambient Concentrations Between the Reference Case and the Control Case in 2018: Percent Changes (left) and Absolute Changes in μg/m³ (right)

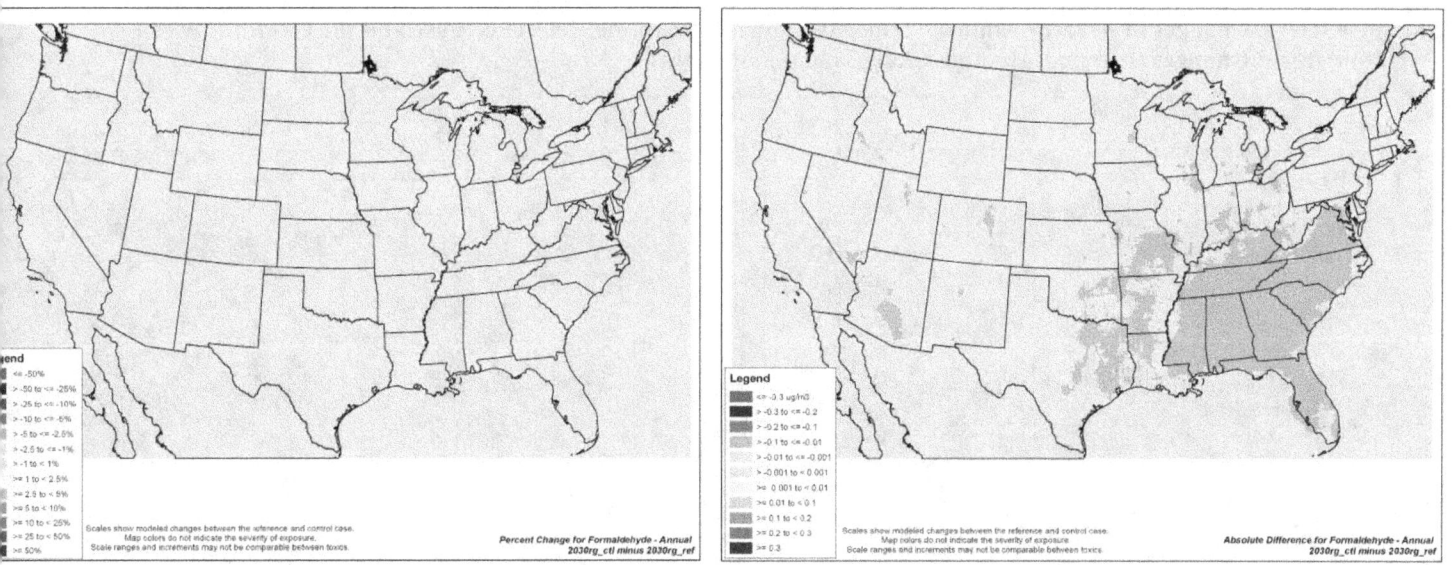

Figure III-12. Changes in Formaldehyde Ambient Concentrations Between the Reference Case and the Control Case in 2030: Percent Changes (left) and Absolute Changes in μg/m³ (right)

3. Benzene

Our air quality modeling projects that the proposed standards would have a notable impact on ambient benzene concentrations. In 2018, soon after the Tier 3 standards take effect, ambient benzene reductions are generally between 0.001 and 0.01 μg/m³, or between 1 and 2.5 percent in some areas (Figure III-13). In 2030, our modeling projects that the proposal would decrease ambient benzene concentrations across much of the country on the order of 1 to 5 percent, with reductions ranging from 10 to 25 percent in some urban areas (Figure III-14).

21

Absolute decreases in ambient concentrations of benzene are generally between 0.001 and 0.01 µg/m³ in rural areas and as much as 0.1 µg/m³ in urban areas (Figure III-14).

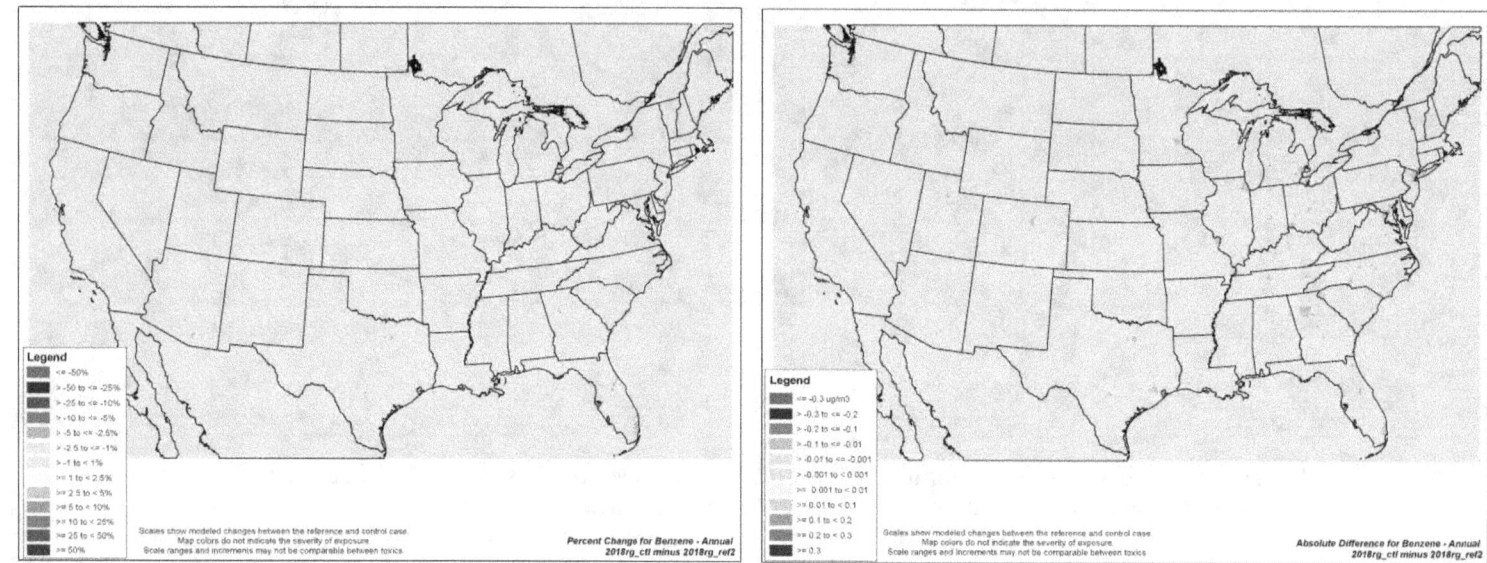

Figure III-13. Changes in Benzene Ambient Concentrations Between the Reference Case and the Control Case in 2018: Percent Changes (left) and Absolute Changes in µg/m³ (right)

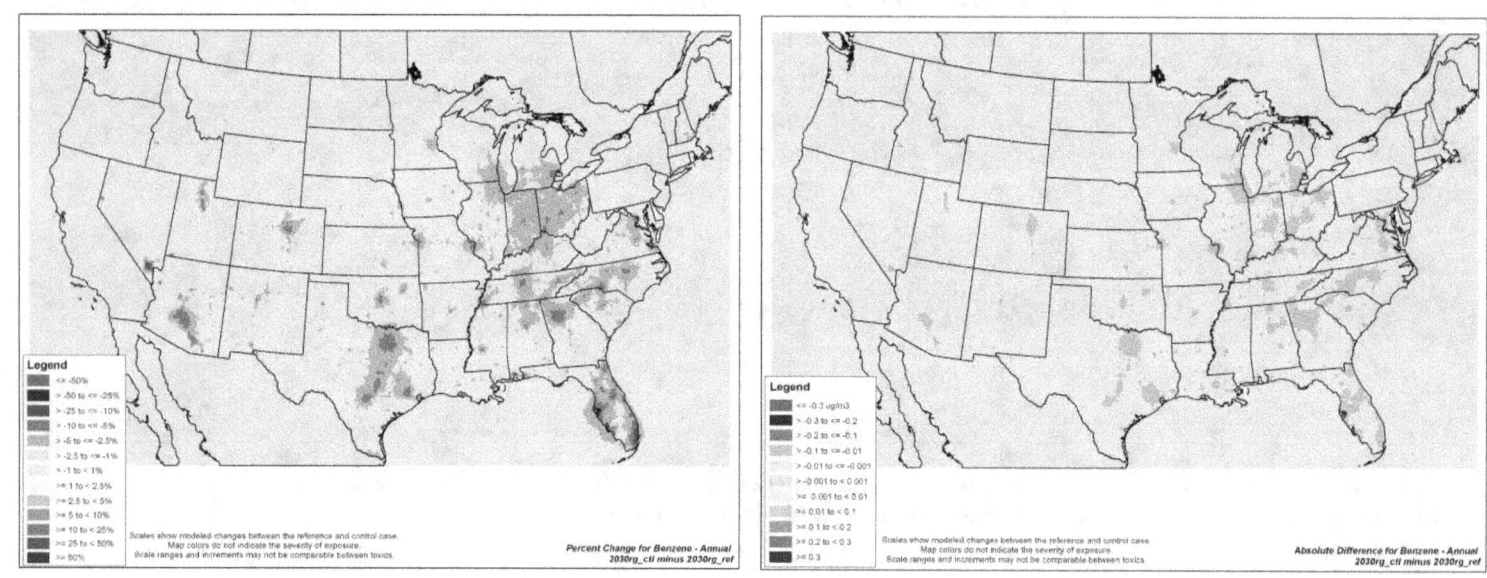

Figure III-14. Changes in Benzene Ambient Concentrations Between the Reference Case and the Control Case in 2030: Percent Changes (left) and Absolute Changes in µg/m³ (right)

4. 1,3-Butadiene

Our modeling also shows reductions of ambient 1,3-butadiene concentrations in 2018 and 2030. Figure III-15 shows that in 2018, ambient concentrations of 1,3-butadiene generally decrease between 1 and 5 percent across the country, corresponding to small decreases in absolute concentrations (less than 0.001 µg/m³). In 2030, reductions of 1,3-butadiene concentrations range between 1 and 25 percent, with decreases of at least 0.005 µg/m³ in urban areas (Figure III-16).

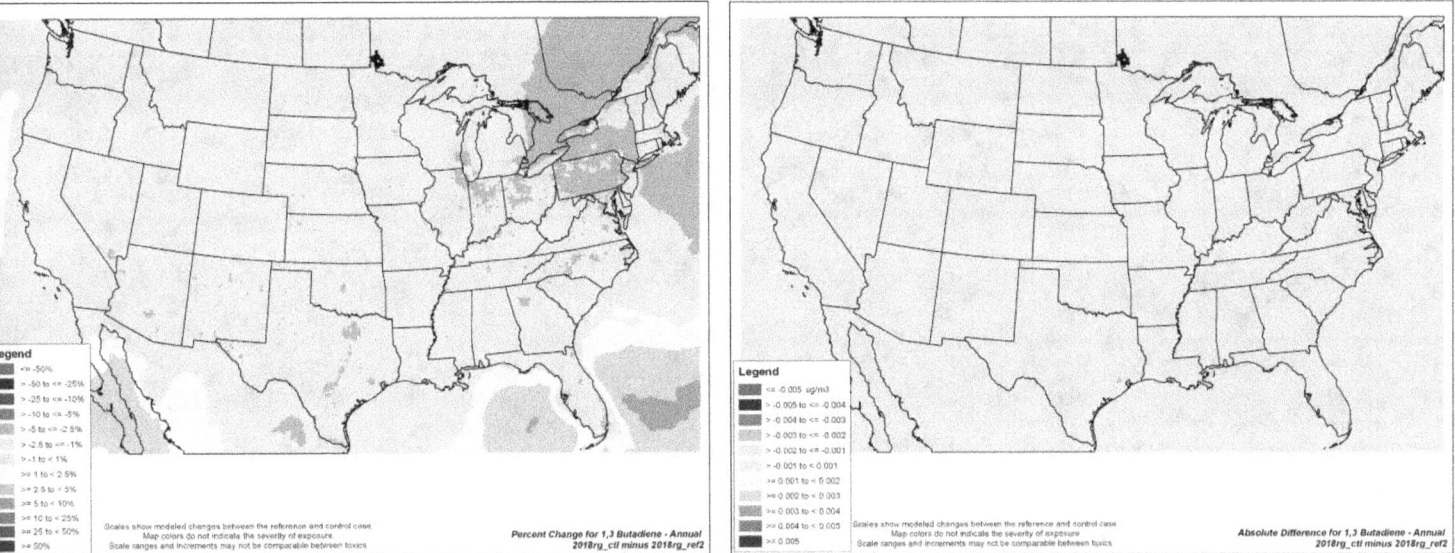

Figure III-15. Changes in 1,3-Butadiene Ambient Concentrations Between the Reference Case and the Control Case in 2018: Percent Changes (left) and Absolute Changes in µg/m³ (right)

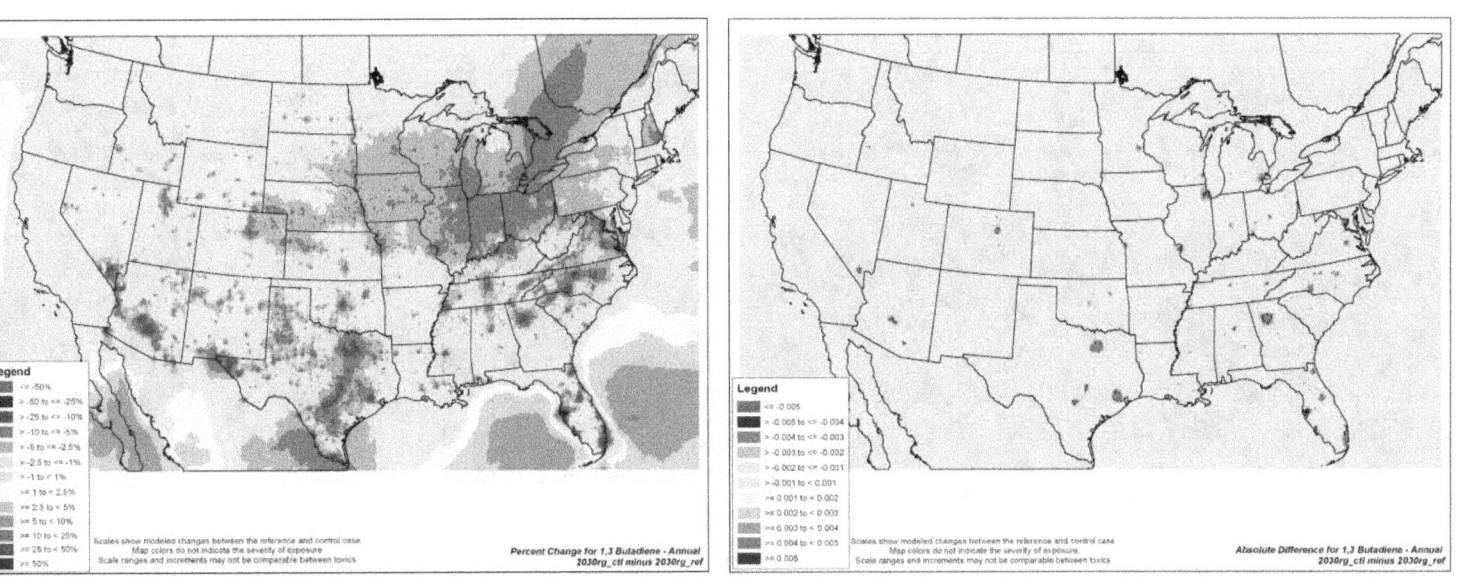

Figure III-16. Changes in 1,3-Butadiene Ambient Concentrations Between the Reference Case and the Control Case in 2030: Percent Changes (left) and Absolute Changes in µg/m³ (right)

5. Acrolein

Our modeling indicates the proposed standards would reduce ambient concentrations of acrolein in 2018 and 2030. Figure III-17 shows decreases in ambient concentrations of acrolein generally between 1 and 2.5 percent across the parts of the country in 2018, corresponding to small decreases in absolute concentrations (less than 0.001 µg/m³). Reductions of acrolein concentrations in 2030 range between 1 and 25 percent, with decreases as high as 0.003 µg/m³ in a few urban areas (Figure III-18).

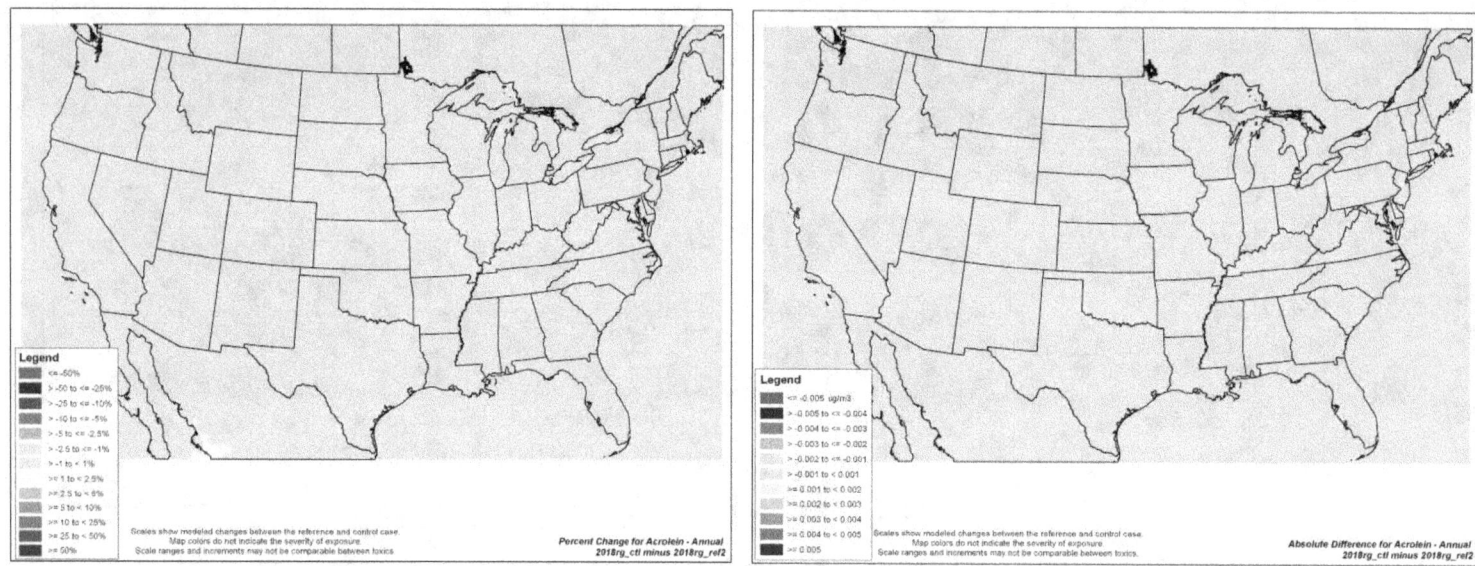

Figure III-17. Changes in Acrolein Ambient Concentrations Between the Reference Case and the Control Case in 2018: Percent Changes (left) and Absolute Changes in µg/m³ (right)

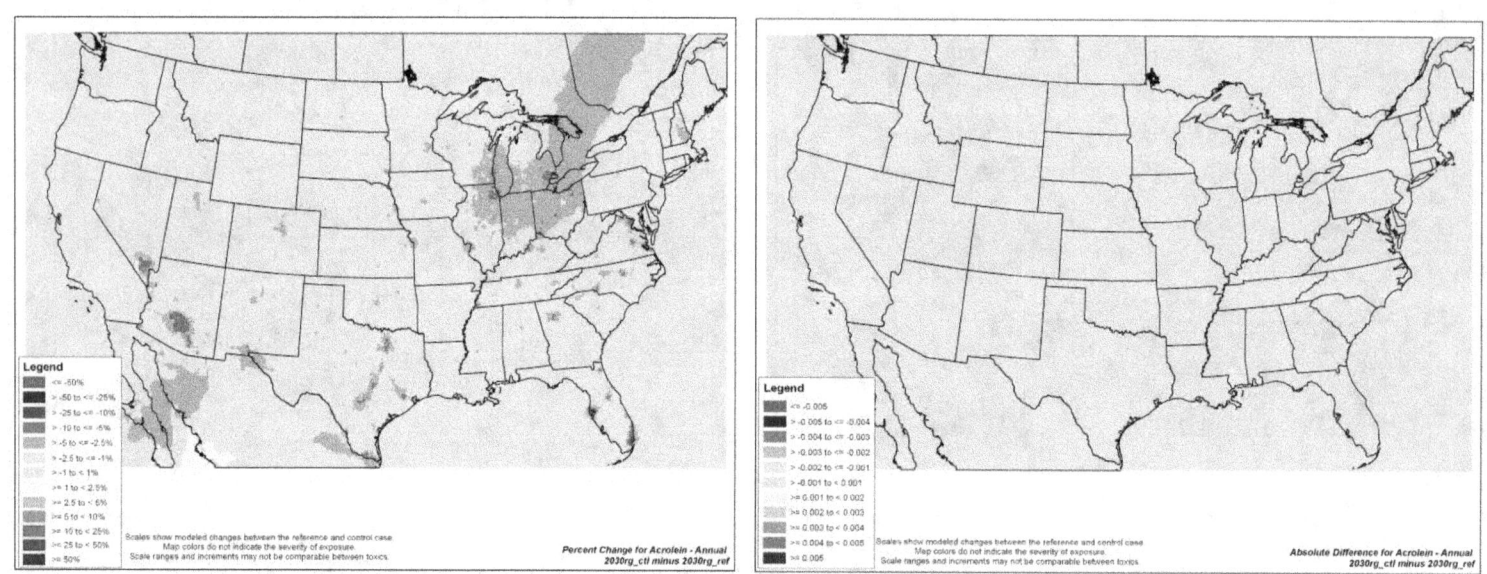

Figure III-18. Changes in Acrolein Ambient Concentrations Between the Reference Case and the Control Case in 2030: Percent Changes (left) and Absolute Changes in µg/m³ (right)

24

6. Ethanol

Our modeling projects that the proposed standards would slightly decrease ambient ethanol concentrations in 2018 and 2030. As shown in Figure III-19, in 2018, annual percent changes in ambient concentrations of ethanol are less than 1 percent across the country, with absolute concentrations of up to 0.1 ppb in some places. In 2030, some parts of the country, especially urban areas, are projected to have reductions in ethanol concentrations on the order of 1 to 10 percent as a result of the rule (Figure III-20). Figure III-20 also shows that absolute decreases in ambient concentrations of ethanol are generally between 0.001 and 0.1 ppb in 2030 with decreases in a few urban areas as high as 0.2 ppb.

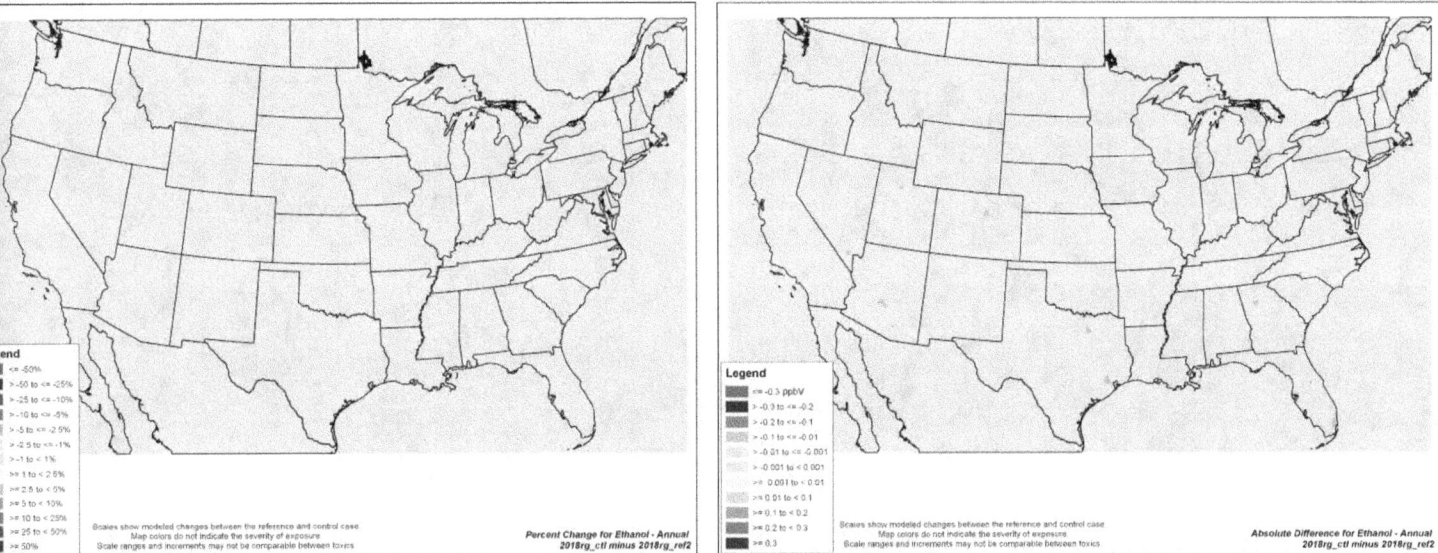

Figure III-19. Changes in Ethanol Ambient Concentrations Between the Reference Case and the Control Case in 2018: Percent Changes (left) and Absolute Changes in µg/m³ (right)

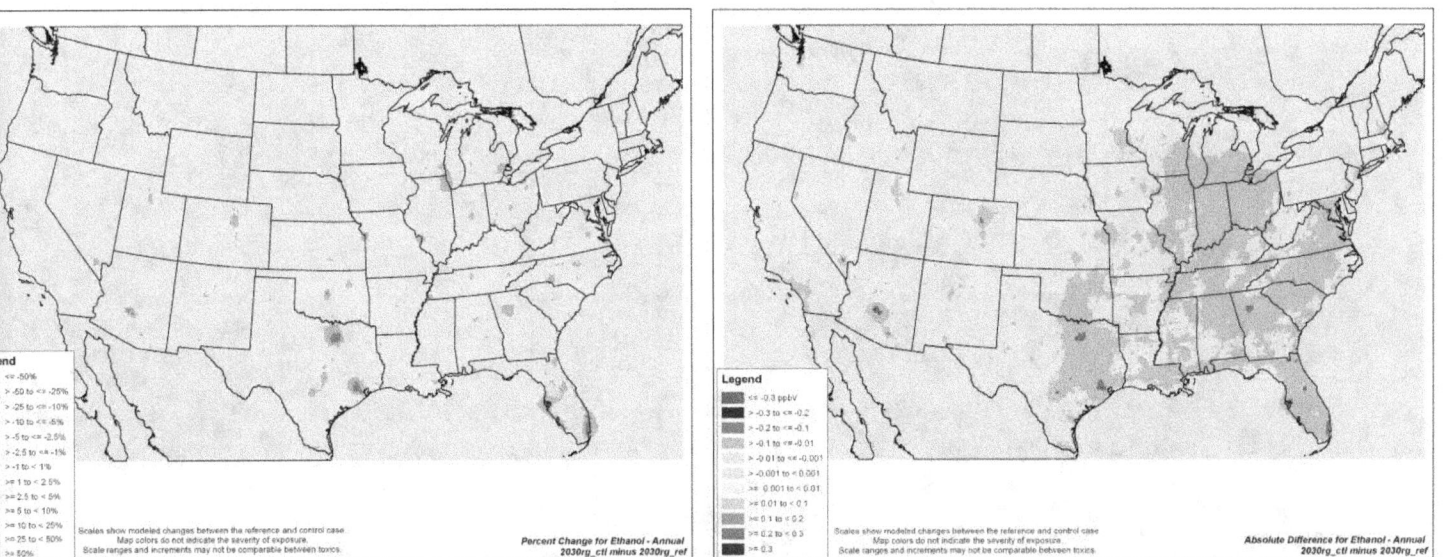

Figure III-20. Changes in Ethanol Ambient Concentrations Between the Reference Case and the Control Case in 2030: Percent Changes (left) and Absolute Changes in µg/m³ (right)

25

7. Naphthalene

Our modeling projects reductions in naphthalene concentrations in 2018 and 2030. As shown in Figure III-21 and Figure III-22, annual percent changes in ambient concentrations of naphthalene are between 1 and 2.5 percent across much of the country for 2018, with small decreases in absolute concentrations (less than 0.001 µg/m³). In 2030, reductions of naphthalene concentrations generally range between 1 and 10 percent but are as high as 25 percent in some areas of the Southeast, with corresponding absolute decreases in urban areas of up to 0.005 µg/m³.

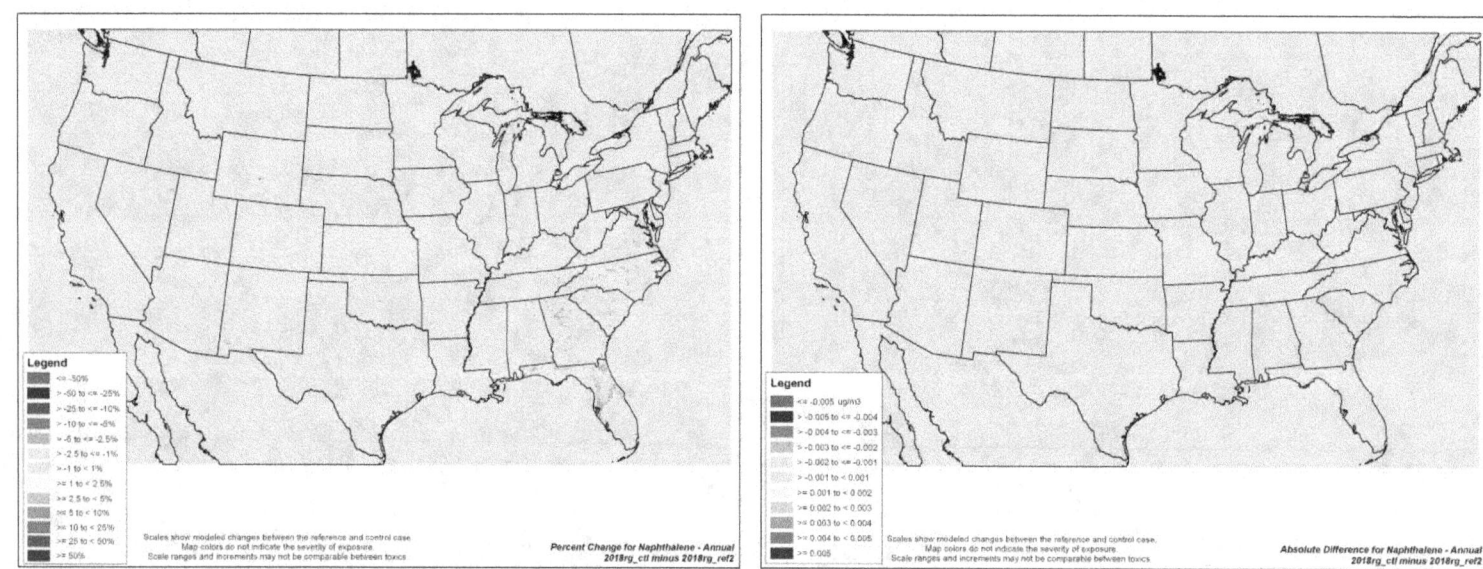

Figure III-21. Changes in Naphthalene Ambient Concentrations Between the Reference Case and the Control Case in 2018: Percent Changes (left) and Absolute Changes in µg/m³ (right)

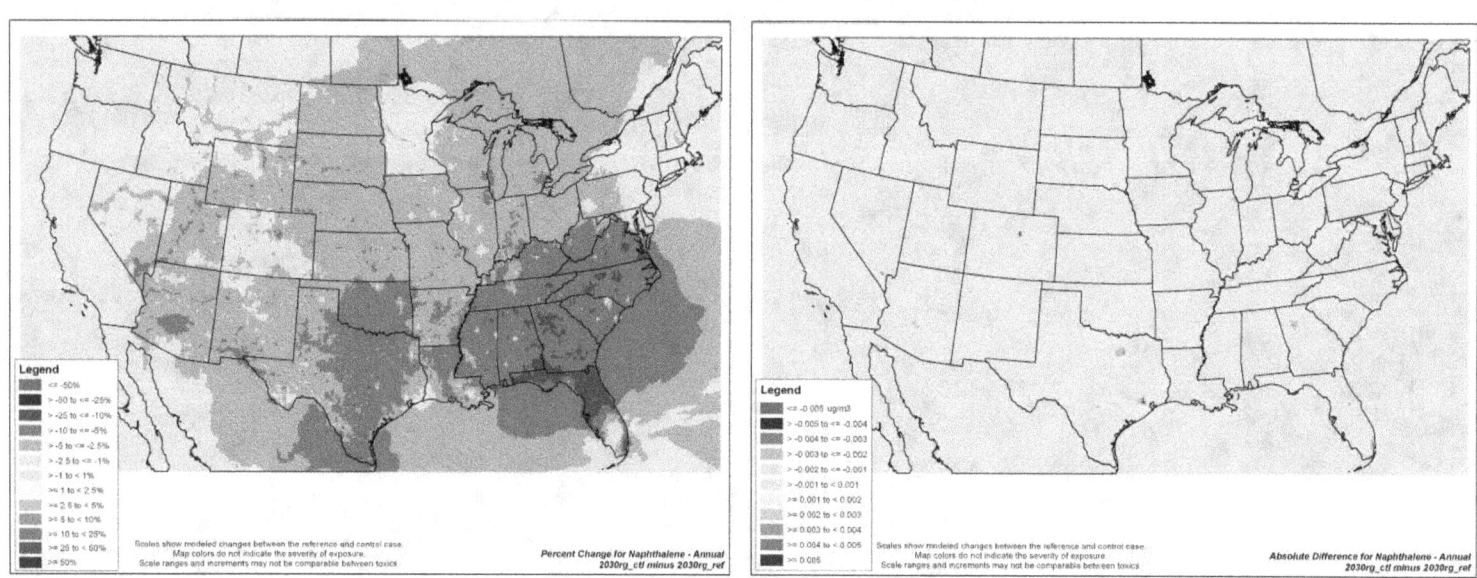

Figure III-22. Changes in Naphthalene Ambient Concentrations Between the Reference Case and the Control Case in 2030: Percent Changes (left) and Absolute Changes in µg/m³ (right)

F. Air Toxics Population Metrics

To assess the impact of the Tier 3 rule on projected changes in air quality, we developed population metrics that show population experiencing changes in annual ambient concentrations across the modeled air toxics. Although the reductions in ambient air toxics concentrations expected from the Tier 3 standards are generally small, they are projected to benefit the majority of the U.S. population. As shown in

Table III-1, over 75 percent of the total U.S. population is projected to experience a decrease in ambient benzene and 1,3-butadiene concentrations of at least 1 percent.

Table III-1 also shows that over 60 percent of the U.S population is projected to experience at least a 1 percent decrease in ambient ethanol and acrolein concentrations, and over 35 percent would experience a similar decrease in ambient formaldehyde concentrations with the standards.

Table III-1. Percent of Total Population Experiencing Changes in Annual Ambient Concentrations of Toxic Pollutants in 2030 as a Result of the Tier 3 Standards

rcent Change	Benzene	Acrolein	1,3-Butadiene	Formaldehyde	Ethanol	Acetaldehyde	Naphthalene
≤ -50							
-50 to $\leq$ -25							
-25 to $\leq$ -10	2.29%	0.75%	19.07%				10.74%
-10 to $\leq$ -5	20.63%	12.72%	27.29%		5.39%		31.56%
-5 to $\leq$ -2.5	27.50%	25.17%	15.37%	0.60%	24.08%		20.58%
-2.5 to $\leq$ -1	28.60%	24.62%	18.33%	35.34%	34.10%	11.77%	14.98%
> -1 to < 1	20.97%	36.74%	19.93%	64.06%	36.43%	88.23%	22.14%
$\geq$ 1 to < 2.5							
$\geq$ 2.5 to < 5							
$\geq$ 5 to < 10							
$\geq$ 10 to < 25							
$\geq$ 25 to < 50							
$\geq$ 50							

G. Impacts of Tier 3 Standards on Future Annual Nitrogen and Sulfur Deposition Levels

Our air quality modeling projects decreases in both nitrogen and sulfur deposition due to this rule. Figure III-23 shows that for nitrogen deposition by 2030 the proposed standards would result in annual percent decreases of more than 2.5 percent in most urban areas with decreases of more than 5 percent in urban areas in Nevada, Florida, Georgia and Virginia. In addition, smaller decreases, in the 1 to 1.5 percent range, would occur over most of the rest of the country.

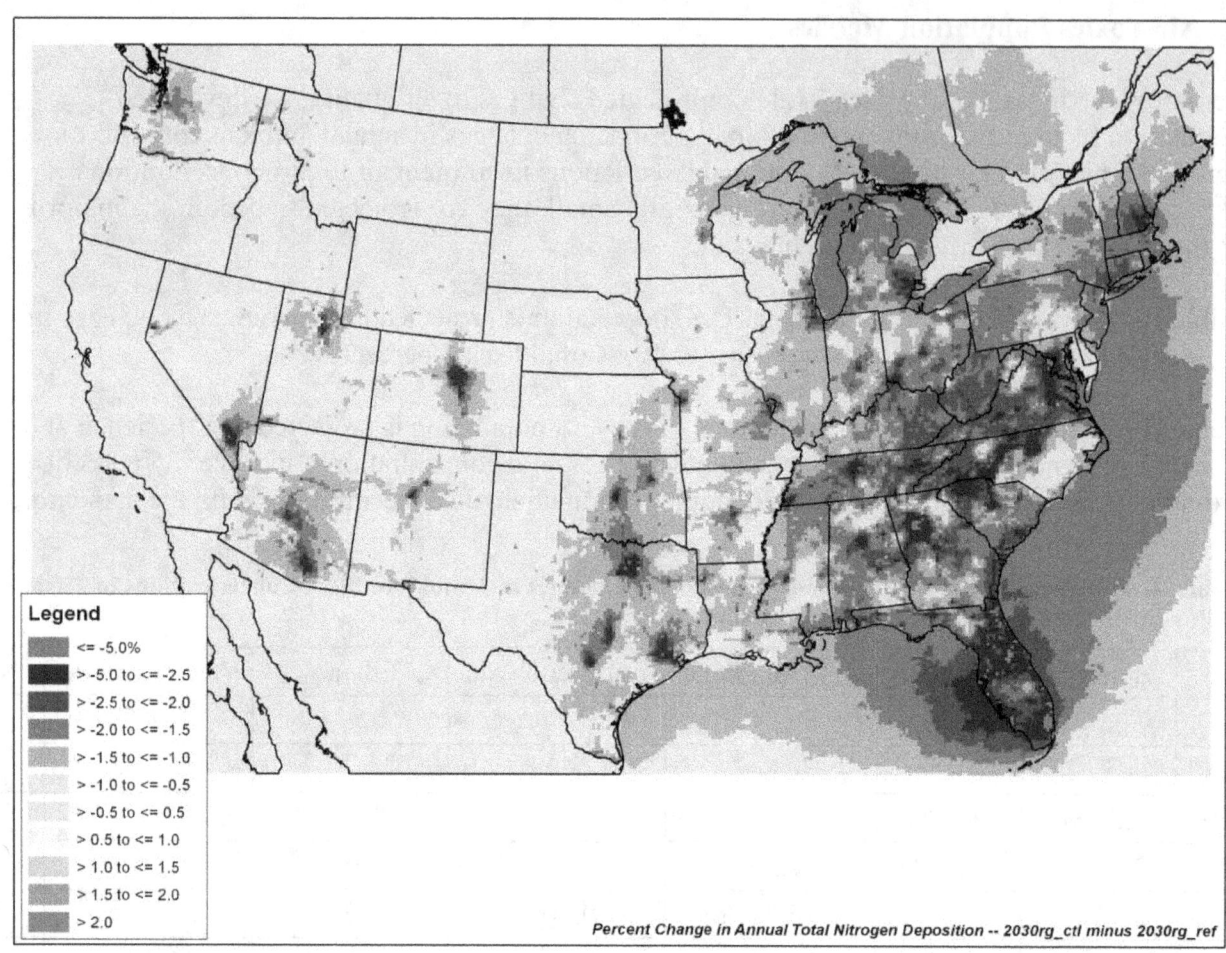

Figure III-23. Percent Change in Annual Total Nitrogen Deposition over the U.S. Modeling Domain as a Result of the Tier 3 Standards in 2030

Figure III-24 shows that for sulfur deposition the Tier 3 standards will result in annual percent decreases of more than 2 percent in some urban areas in 2030. The decreases in sulfur deposition are likely due to projected reductions in the sulfur level in fuel. Minimal changes in sulfur deposition, ranging from decreases of less than 0.5 percent to no change, are projected for the rest of the country.

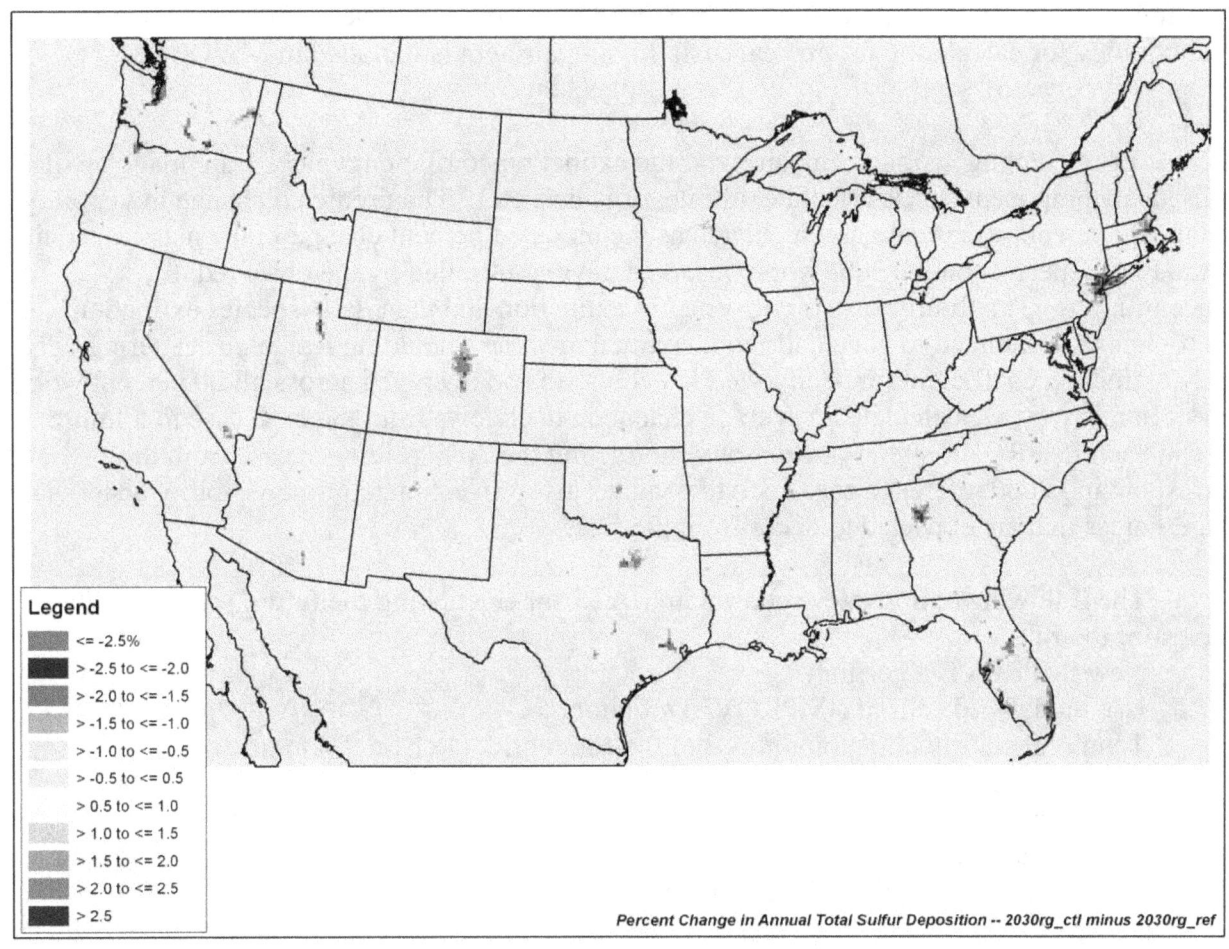

Legend
- <= -2.5%
- > -2.5 to <= -2.0
- > -2.0 to <= -1.5
- > -1.5 to <= -1.0
- > -1.0 to <= -0.5
- > -0.5 to <= 0.5
- > 0.5 to <= 1.0
- > 1.0 to <= 1.5
- > 1.5 to <= 2.0
- > 2.0 to <= 2.5
- > 2.5

Percent Change in Annual Total Sulfur Deposition -- 2030rg_ctl minus 2030rg_ref

Figure III-24. Percent Changes in Annual Total Sulfur Deposition over the U.S. Modeling Domain as a Result of the Tier 3 Standards in 2030

H. Impacts of Tier 3 Standards on Future Visibility Levels

Air quality modeling conducted for the Tier 3 rule was used to project visibility conditions in 137 Mandatory Class I Federal areas across the U.S. in 2018 and 2030. The impacts of this action were examined in terms of the projected improvements in visibility on the 20 percent worst visibility days at Class I areas. We quantified visibility impacts at the Class I areas which have complete IMPROVE ambient data for 2007 or are represented by IMPROVE monitors with complete data. Sites were used in this analysis if they had at least 3 years of complete data for the 2005-2009 period[30].

Visibility for the 2018 and 2030 reference and control cases were calculated using the regional haze methodology outlined in section 6 of the photochemical modeling guidance, which applies modeling results in a relative sense, using base year ambient data. The $PM_{2.5}$ and

---

[30] Since the base case modeling used meteorology for 2007, one of the complete years must be 2007.

regional haze modeling guidance recommends the calculation of future year changes in visibility in a similar manner to the calculation of changes in $PM_{2.5}$ design values. The regional haze methodology for calculating future year visibility impairment is included in MATS (http://www.epa.gov/scram001/modelingapps_mats.htm)

In calculating visibility impairment, the extinction coefficient values[31] are made up of individual component species (sulfate, nitrate, organics, etc). The predicted change in visibility (on the 20 percent worst days) is calculated as the modeled percent change in the mass for each of the $PM_{2.5}$ species (on the 20% worst observed days) multiplied by the observed concentrations. The future mass is converted to extinction and then daily species extinction coefficients are summed to get a daily total extinction value (including Rayleigh scattering). The daily extinction coefficients are converted to deciviews and averaged across all 20 percent worst days. In this way, we calculate an average change in deciviews from the base case to a future case at each IMPROVE site. For example, subtracting the 2030 reference case from the corresponding 2030 reference case deciview values gives an estimate of the visibility benefits in Class I areas that are expected to occur from the rule.

The following options were chosen in MATS for calculating the future year visibility values for the rule:
New IMPROVE algorithm
Use model grid cells at (IMPROVE) monitor
Temporal adjustment at monitor- 3x3 for 12km grid, (1x1 for 36km grid)
Start monitor year- 2005
End monitor year- 2009
Base model year 2007
Minimum years required for a valid monitor- 3

The "base model year" was chosen as 2007 because it is the base case meteorological year for the Tier 3 final rule modeling. The start and end years were chosen as 2005 and 2009 because that is the 5 year period which is centered on the base model year of 2007. These choices are consistent with using a 5 year base period for regional haze calculations.

The results show that in 2030 all the modeled areas would continue to have annual average deciview levels above background and the rule would improve visibility in all these areas.[32] The average visibility on the 20 percent worst days at all modeled Mandatory Class I Federal areas is projected to improve by 0.02 deciviews, or 0.16 percent, in 2030. The greatest improvement in visibilities will be seen in Craters of the Moon National Monument, where visibility is projected to improve by 0.7 percent (0.09 DV) in 2030 due to the standards. Table III-2 contains the full visibility results from 2018 and 2030 for the 137 analyzed areas.

---

[31] Extinction coefficient is in units of inverse megameters (Mm^{-1}). It is a measure of how much light is absorbed or scattered as it passes through a medium. Light extinction is commonly used as a measure of visibility impairment in the regional haze program.

[32] The level of visibility impairment in an area is based on the light-extinction coefficient and a unitless visibility index, called a "deciview", which is used in the valuation of visibility. The deciview metric provides a scale for perceived visual changes over the entire range of conditions, from clear to hazy. Under many scenic conditions, the average person can generally perceive a change of one deciview. The higher the deciview value, the worse the visibility. Thus, an improvement in visibility is a decrease in deciview value.

Table III-2. Visibility Levels in Deciviews for Mandatory Class I Federal Areas on the 20 Percent Worst Days with and without Tier 3 Rule

Class 1 Area (20% worst days)	State	2007 Baseline Visibility (dv)[a]	2018 Reference	2018 Tier 3 Control	2030 Reference	2030 Tier 3 Control	Natural Background
Sipsey Wilderness	AL	28.32	20.59	20.55	20.43	20.37	10.99
Upper Buffalo Wilderness	AR	25.86	20.01	19.98	19.93	19.88	11.57
Chiricahua NM	AZ	12.22	11.82	11.82	12.38	12.37	7.20
Chiricahua Wilderness	AZ	12.22	11.83	11.82	12.38	12.37	7.20
Galiuro Wilderness	AZ	12.22	11.99	11.98	12.41	12.40	7.20
Grand Canyon NP	AZ	11.97	11.21	11.20	11.31	11.30	7.04
Mazatzal Wilderness	AZ	13.40	12.65	12.65	12.88	12.85	6.68
Mount Baldy Wilderness	AZ	11.79	10.98	10.98	11.24	11.22	6.24
Petrified Forest NP	AZ	13.02	12.24	12.23	12.37	12.35	6.49
Pine Mountain Wilderness	AZ	13.40	12.69	12.69	12.93	12.91	6.68
Saguaro NM	AZ	13.63	13.02	13.00	13.04	12.99	6.46
Superstition Wilderness	AZ	13.81	13.18	13.18	13.38	13.34	6.54
Sycamore Canyon Wilderness	AZ	15.18	14.94	14.94	15.03	15.02	6.65
Agua Tibia Wilderness	CA	20.92	17.67	17.66	16.85	16.85	7.64
Ansel Adams Wilderness (Minarets)	CA	15.72	14.57	14.57	14.38	14.38	7.12
Caribou Wilderness	CA	15.99	15.54	15.54	15.48	15.48	7.31
Cucamonga Wilderness	CA	18.03	15.37	15.36	14.91	14.90	6.99
Desolation Wilderness	CA	13.62	12.89	12.89	12.76	12.75	6.05
Dome Land Wilderness	CA	19.23	17.89	17.89	17.60	17.60	7.46
Emigrant Wilderness	CA	16.87	15.84	15.84	15.67	15.66	7.64
Hoover Wilderness	CA	12.19	11.49	11.48	11.41	11.41	7.71
John Muir Wilderness	CA	15.72	14.76	14.76	14.60	14.60	7.12
Joshua Tree NM	CA	17.83	15.75	15.75	15.33	15.32	7.19
Kaiser Wilderness	CA	15.72	14.80	14.80	14.59	14.59	7.12
Kings Canyon NP	CA	23.39	21.56	21.55	21.06	21.05	7.70
Lassen Volcanic NP	CA	15.99	15.52	15.52	15.45	15.45	7.31
Lava Beds NM	CA	14.17	13.78	13.78	13.68	13.67	7.85
Marble Mountain Wilderness	CA	17.34	17.02	17.01	16.91	16.91	7.90
Mokelumne Wilderness	CA	13.62	12.88	12.88	12.75	12.75	6.05
Pinnacles NM	CA	18.37	16.44	16.43	16.05	16.05	7.99
Point Reyes NS	CA	22.03	21.04	21.03	20.71	20.71	15.77
Redwood NP	CA	19.14	18.72	18.70	18.43	18.42	13.91
San Gabriel Wilderness	CA	18.03	15.71	15.71	15.31	15.30	6.99
San Gorgonio Wilderness	CA	20.48	17.68	17.68	16.94	16.93	7.30
San Jacinto Wilderness	CA	20.48	17.76	17.76	16.95	16.95	7.30
San Rafael Wilderness	CA	19.20	17.46	17.46	17.10	17.10	7.57
Sequoia NP	CA	23.39	21.28	21.28	20.74	20.73	7.70
South Warner Wilderness	CA	14.17	13.60	13.60	13.49	13.49	7.85

Class 1 Area (20% worst days)	State	2007 Baseline Visibility (dv)[a]	2018 Reference	2018 Tier 3 Control	2030 Reference	2030 Tier 3 Control	Natural Background
Thousand Lakes Wilderness	CA	15.99	15.53	15.53	15.46	15.45	7.31
Ventana Wilderness	CA	18.37	16.79	16.79	16.50	16.49	7.99
Yolla Bolly Middle Eel Wilderness	CA	17.34	17.06	17.06	16.99	16.99	7.90
Yosemite NP	CA	16.87	15.98	15.98	15.85	15.84	7.64
Black Canyon of the Gunnison NM	CO	10.04	9.21	9.20	9.26	9.24	6.21
Eagles Nest Wilderness	CO	8.94	7.98	7.97	7.97	7.93	6.06
Flat Tops Wilderness	CO	8.94	8.26	8.26	8.28	8.27	6.06
Great Sand Dunes NM	CO	11.44	10.57	10.56	10.59	10.57	6.66
La Garita Wilderness	CO	10.04	9.36	9.35	9.44	9.43	6.21
Maroon Bells-Snowmass Wilderness	CO	8.94	8.15	8.14	8.18	8.17	6.06
Mesa Verde NP	CO	11.28	10.48	10.47	10.57	10.55	6.81
Mount Zirkel Wilderness	CO	9.72	9.12	9.11	9.10	9.08	6.08
Rawah Wilderness	CO	9.72	8.92	8.91	8.88	8.86	6.08
Rocky Mountain NP	CO	12.62	11.66	11.64	11.55	11.50	7.15
Weminuche Wilderness	CO	10.04	9.38	9.37	9.45	9.44	6.21
West Elk Wilderness	CO	8.94	8.12	8.11	8.18	8.16	6.06
Chassahowitzka	FL	23.68	18.63	18.59	18.38	18.31	11.03
Everglades NP	FL	20.41	17.43	17.42	17.28	17.25	12.15
St. Marks	FL	25.58	20.07	20.04	19.86	19.81	11.67
Cohutta Wilderness	GA	28.01	18.77	18.73	18.59	18.52	10.78
Okefenokee	GA	26.00	21.32	21.30	21.33	21.31	11.44
Wolf Island	GA	26.00	20.53	20.51	20.45	20.41	11.44
Craters of the Moon NM	ID	13.63	12.91	12.86	12.63	12.54	7.53
Sawtooth Wilderness	ID	14.76	14.61	14.61	14.58	14.57	6.42
Mammoth Cave NP	KY	30.68	21.59	21.55	21.47	21.41	11.08
Acadia NP	ME	21.45	17.41	17.38	17.22	17.19	12.43
Moosehorn	ME	19.92	16.23	16.21	16.14	16.12	12.01
Roosevelt Campobello International Park	ME	19.92	16.45	16.43	16.34	16.32	12.01
Isle Royale NP	MI	21.76	18.49	18.45	18.21	18.13	12.37
Seney	MI	24.21	20.30	20.26	20.17	20.09	12.65
Boundary Waters Canoe Area	MN	20.05	17.05	17.01	16.77	16.70	11.61
Voyageurs NP	MN	19.78	17.60	17.57	17.35	17.29	12.06
Hercules-Glades Wilderness	MO	26.05	20.36	20.32	20.21	20.14	11.30
Mingo	MO	27.08	21.09	21.06	20.88	20.83	11.62
Bob Marshall Wilderness	MT	15.32	15.13	15.13	15.06	15.05	7.73
Cabinet Mountains Wilderness	MT	13.47	13.16	13.15	13.01	13.00	7.52
Glacier NP	MT	18.70	18.39	18.38	18.23	18.21	9.18
Medicine Lake	MT	18.02	16.67	16.66	16.47	16.45	7.89

Class 1 Area (20% worst days)	State	2007 Baseline Visibility (dv)[a]	2018 Reference	2018 Tier 3 Control	2030 Reference	2030 Tier 3 Control	Natural Background
Mission Mountains Wilderness	MT	15.32	15.08	15.07	14.98	14.97	7.73
Red Rock Lakes	MT	11.53	11.20	11.19	11.13	11.11	6.44
Scapegoat Wilderness	MT	15.32	15.17	15.17	15.12	15.11	7.73
UL Bend	MT	14.86	14.41	14.41	14.37	14.36	8.16
Linville Gorge Wilderness	NC	27.39	18.40	18.37	18.33	18.28	11.22
Shining Rock Wilderness	NC	26.60	18.17	18.13	18.04	17.98	11.47
Lostwood	ND	19.56	18.58	18.57	18.45	18.44	8.00
Great Gulf Wilderness	NH	20.19	15.15	15.13	15.08	15.05	11.99
Presidential Range-Dry River Wilderness	NH	20.19	15.05	15.03	14.97	14.94	11.99
Brigantine	NJ	27.32	20.66	20.63	20.59	20.55	12.24
Bandelier NM	NM	11.84	10.81	10.79	10.89	10.85	6.26
Bosque del Apache	NM	13.40	12.32	12.30	12.54	12.50	6.73
Carlsbad Caverns NP	NM	15.85	15.19	15.18	15.88	15.86	6.65
Gila Wilderness	NM	12.49	11.94	11.94	12.40	12.39	6.66
Pecos Wilderness	NM	9.13	8.19	8.18	8.34	8.32	6.08
San Pedro Parks Wilderness	NM	9.89	9.06	9.05	9.28	9.27	5.72
Wheeler Peak Wilderness	NM	9.13	8.13	8.13	8.25	8.23	6.08
White Mountain Wilderness	NM	13.20	12.34	12.33	12.74	12.73	6.80
Jarbidge Wilderness	NV	12.42	12.17	12.16	12.13	12.12	7.87
Wichita Mountains	OK	22.97	19.63	19.60	19.52	19.45	7.53
Crater Lake NP	OR	13.79	13.33	13.32	13.22	13.22	7.62
Diamond Peak Wilderness	OR	13.79	13.23	13.22	13.07	13.07	7.62
Eagle Cap Wilderness	OR	16.23	15.61	15.59	15.22	15.20	8.92
Gearhart Mountain Wilderness	OR	13.79	13.35	13.35	13.27	13.27	7.62
Hells Canyon Wilderness	OR	18.15	17.54	17.50	17.20	17.16	8.32
Kalmiopsis Wilderness	OR	16.45	15.82	15.81	15.63	15.62	9.44
Mount Hood Wilderness	OR	13.72	12.71	12.68	12.25	12.23	8.43
Mount Jefferson Wilderness	OR	16.18	15.58	15.57	15.33	15.31	8.79
Mount Washington Wilderness	OR	16.18	15.57	15.55	15.32	15.31	8.79
Mountain Lakes Wilderness	OR	13.79	13.28	13.28	13.16	13.16	7.62
Strawberry Mountain Wilderness	OR	16.23	15.37	15.34	15.00	14.97	8.92
Three Sisters Wilderness	OR	16.18	15.63	15.61	15.45	15.44	8.79
Cape Romain	SC	26.45	19.75	19.72	19.61	19.56	12.12
Badlands NP	SD	16.55	15.25	15.24	15.19	15.17	8.06
Wind Cave NP	SD	15.50	14.41	14.39	14.26	14.24	7.71
Great Smoky Mountains NP	TN	28.50	19.57	19.52	19.44	19.38	11.24

Class 1 Area (20% worst days)	State	2007 Baseline Visibility (dv)[a]	2018 Reference	2018 Tier 3 Control	2030 Reference	2030 Tier 3 Control	Natural Background
Joyce-Kilmer-Slickrock Wilderness	TN	28.50	19.65	19.61	19.52	19.46	11.24
Big Bend NP	TX	16.69	16.39	16.38	17.32	17.31	7.16
Guadalupe Mountains NP	TX	15.85	15.23	15.22	15.94	15.92	6.65
Arches NP	UT	11.02	10.33	10.32	10.30	10.27	6.43
Bryce Canyon NP	UT	11.88	11.40	11.40	11.39	11.37	6.80
Canyonlands NP	UT	11.02	10.50	10.48	10.57	10.55	6.43
Capitol Reef NP	UT	11.30	10.73	10.72	10.74	10.72	6.03
James River Face Wilderness	VA	27.29	19.05	19.02	18.89	18.83	11.13
Shenandoah NP	VA	27.26	17.67	17.63	17.60	17.54	11.35
Lye Brook Wilderness	VT	23.01	16.74	16.70	16.58	16.53	11.73
Alpine Lake Wilderness	WA	16.09	14.87	14.84	14.22	14.17	8.43
Glacier Peak Wilderness	WA	13.72	12.78	12.77	12.56	12.54	8.39
Goat Rocks Wilderness	WA	12.66	11.92	11.90	11.66	11.64	8.35
Mount Adams Wilderness	WA	12.66	12.04	12.02	11.77	11.75	8.35
Mount Rainier NP	WA	16.38	15.53	15.52	15.25	15.24	8.54
North Cascades NP	WA	13.72	12.87	12.86	12.71	12.70	8.01
Olympic NP	WA	15.20	14.30	14.28	13.94	13.92	8.44
Pasayten Wilderness	WA	14.09	13.51	13.50	13.26	13.25	8.25
Dolly Sods Wilderness	WV	27.55	17.97	17.94	17.99	17.95	10.39
Otter Creek Wilderness	WV	27.55	18.11	18.07	18.08	18.04	10.39
Bridger Wilderness	WY	10.68	10.23	10.22	10.20	10.19	6.45
Fitzpatrick Wilderness	WY	10.68	10.21	10.21	10.18	10.17	6.45
Grand Teton NP	WY	11.53	11.14	11.13	11.09	11.07	6.44
Teton Wilderness	WY	11.53	11.18	11.18	11.15	11.14	6.44
Yellowstone NP	WY	11.53	11.26	11.26	11.23	11.22	6.44

Air Quality Modeling Technical Support Document: Tier 3 Motor Vehicle Emission and Fuel Standards

Appendix A

Model Performance Evaluation for the 2007-Based Air Quality Modeling Platform

U.S. Environmental Protection Agency
Office of Air Quality Planning and Standards
Air Quality Assessment Division
Research Triangle Park, NC 27711
February 2014

A.1. Introduction

An operational model performance evaluation for ozone, $PM_{2.5}$ and its related speciated components, specific air toxics (i.e., formaldehyde, acetaldehyde, benzene, 1,3-butadiene, and acrolein), as well as nitrate and sulfate deposition was conducted using 2007 State/local monitoring sites data in order to estimate the ability of the CMAQ modeling system to replicate the base year concentrations for the 12 km Continental United States domain[1]. Included in this evaluation are statistical measures of model versus observed pairs that were paired in space and time on a daily or weekly basis, depending on the sampling frequency of each network (measured data). For certain time periods with missing ozone, $PM_{2.5}$, air toxic observations and nitrate and sulfate deposition we excluded the CMAQ predictions from those time periods in our calculations. It should be noted when pairing model and observed data that each CMAQ concentration represents a grid-cell volume-averaged value, while the ambient network measurements are made at specific locations.

Model performance statistics were calculated for several spatial scales and temporal periods. Statistics were generated for five large subregions[2]: Midwest, Northeast, Southeast, Central, and West U.S. The statistics for each site and subregion were calculated by season (e.g., "winter" is defined as December, January, and February). For 8-hour daily maximum ozone, we also calculated performance statistics by subregion for the May through September ozone season[3]. In addition to the performance statistics, we prepared several graphical presentations of model performance. These graphical presentations include:

(1) regional maps which show the normalized mean bias and error calculated for each season at individual monitoring sites, and
(2) bar and whisker plots which show the distribution of the predicted and observed data by month by subregion.

A.1.1 Monitoring Networks

The model evaluation for ozone was based upon comparisons of model predicted 8-hour daily maximum concentrations to the corresponding ambient measurements for 2007 at monitoring sites in the EPA Air Quality System (AQS). The observed ozone data were measured and reported on an hourly basis. The $PM_{2.5}$ evaluation focuses on concentrations of $PM_{2.5}$ total mass and its components including sulfate (SO_4), nitrate (NO_3), total nitrate ($TNO_3=NO_3+HNO_3$), ammonium (NH_4), elemental carbon (EC), and organic carbon (OC) as well as wet deposition for nitrate and sulfate. The $PM_{2.5}$ performance statistics were calculated for each season and for the entire year, as a whole. $PM_{2.5}$ ambient measurements for 2007 were obtained from the following networks: Chemical Speciation Network (CSN), Interagency

---

[1]See section II.B. of the main document (Figure II-1) for the description and map of the CMAQ modeling domains.
[2] The subregions are defined by States where: Midwest is IL, IN, MI, OH, and WI; Northeast is CT, DE, MA, MD, ME, NH, NJ, NY, PA, RI, and VT; Southeast is AL, FL, GA, KY, MS, NC, SC, TN, VA, and WV; Central is AR, IA, KS, LA, MN, MO, NE, OK, and TX; West is AK, CA, OR, WA, AZ, NM, CO, UT, WY, SD, ND, MT, ID, and NV.
[3] In calculating the ozone season statistics we limited the data to those observed and predicted pairs with observations that exceeded 60 ppb in order to focus on concentrations at the upper portion of the distribution of values.

Monitoring of PROtected Visual Environments (IMPROVE), Clean Air Status and Trends Network (CASTNet), and National Acid Deposition Program/National Trends (NADP/NTN). NADP/NTN collects and reports wet deposition measurements as weekly average data. The pollutant species included in the evaluation for each network are listed in Table A-1. For $PM_{2.5}$ species that are measured by more than one network, we calculated separate sets of statistics for each network. The CSN and IMPROVE networks provide 24-hour average concentrations on a 1 in every 3 day, or 1 in every 6 day sampling cycle. The $PM_{2.5}$ species data at CASTNet sites are weekly integrated samples. In this analysis we use the term "urban sites" to refer to CSN sites; "suburban/rural sites" to refer to CASTNet sites; and "rural sites" to refer to IMPROVE sites.

Table A-1. $PM_{2.5}$ monitoring networks and pollutants species included in the CMAQ performance evaluation.

Ambient Monitoring Networks	Particulate Species							Wet Deposition Species	
	$PM_{2.5}$ Mass	SO_4	NO_3	TNO_3[a]	EC	OC	NH_4	SO_4	NO_3
IMPROVE	X	X	X		X	X			
CASTNet		X		X			X		
STN	X	X	X		X	X	X		
NADP								X	X

[a] $TNO_3 = (NO_3 + HNO_3)$

The air toxics evaluation focuses on specific species relevant to the Tier 3 standards and rulemaking, i.e., formaldehyde, acetaldehyde, benzene, 1,3-butadiene, and acrolein. Similar to the $PM_{2.5}$ evaluation, the air toxics performance statistics were calculated for each season and for the entire year, as a whole to estimate the ability of the CMAQ modeling system to replicate the base year concentrations for the 12 km Continental United States domain. As mentioned above, seasons were defined as: winter (December-January-February), spring (March-April-May), summer (June-July-August), and fall (September-October-November). Toxic measurements for 2007 were obtained from the National Air Toxics Trends Stations (NATTS).

A.1.2 Model Performance Statistics

The Atmospheric Model Evaluation Tool (AMET) was used to conduct the evaluation described in this document.[4] There are various statistical metrics available and used by the science community for model performance evaluation. For a robust evaluation, the principal evaluation statistics used to evaluate CMAQ performance were two bias metrics, normalized mean bias and fractional bias; and two error metrics, normalized mean error and fractional error.

---

[4] Appel, K.W., Gilliam, R.C., Davis, N., Zubrow, A., and Howard, S.C.: Overview of the Atmospheric Model Evaluation Tool (AMET) v1.1 for evaluating meteorological and air quality models, *Environ. Modell. Softw.*,26, 4, 434-443, 2011. (http://www.cmascenter.org/)

Normalized mean bias (NMB) is used as a normalization to facilitate a range of concentration magnitudes. This statistic averages the difference (model - observed) over the sum of observed values. NMB is a useful model performance indicator because it avoids over inflating the observed range of values, especially at low concentrations.

Normalized mean bias is defined as:

$$NMB = \frac{\sum_{1}^{n}(P - O)}{\sum_{1}^{n}O} *100$$

Normalized mean error (NME) is also similar to NMB, where the performance statistic is used as a normalization of the mean error. NME calculates the absolute value of the difference (model - observed) over the sum of observed values.

Normalized mean error is defined as:

$$NME = \frac{\sum_{1}^{n}|P - O|}{\sum_{1}^{n}O} *100$$

Fractional bias is defined as:

$$FB = \frac{1}{n}\left(\frac{\sum_{1}^{n}(P - O)}{\sum_{1}^{n}\frac{(P - O)}{2}} \right) *100, \text{ where P = predicted and O = observed concentrations.}$$

FB is a useful model performance indicator because it has the advantage of equally weighting positive and negative bias estimates. The single largest disadvantage in this estimate of model performance is that the estimated concentration (i.e., prediction, P) is found in both the numerator and denominator. Fractional error (FE) is similar to fractional bias except the absolute value of the difference is used so that the error is always positive.

Fractional error is defined as:

$$FE = \frac{1}{n}\left(\frac{\sum_{1}^{n}|P - O|}{\sum_{1}^{n}\frac{(P - O)}{2}} \right) *100$$

The "acceptability" of model performance was judged by comparing our CMAQ 2007 performance results to the range of performance found in recent regional ozone, $PM_{2.5}$, and air toxic model applications.[5,6,7,8,9,10,1112,13,14, 15,16] These other modeling studies represent a wide range of modeling analyses which cover various models, model configurations, domains, years and/or episodes, chemical mechanisms, and aerosol modules. Overall, the ozone, $PM_{2.5}$, air toxics concentrations and nitrate and sulfate deposition model performance results for the 2007 CMAQ simulations performed for the Tier 3 final rule are within the range or close to that found in other recent applications. The model performance results, as described in this report, give us confidence that our applications of CMAQ using this 2007 modeling platform provide a scientifically credible approach for assessing ozone and $PM_{2.5}$ concentrations for the purposes of the Tier 3 final rule.

[5] Appel, K.W., Bhave, P.V., Gilliland, A.B., Sarwar, G., and Roselle, S.J.: evaluation of the community multiscale air quality (CMAQ) model version 4.5: sensitivities impacting model performance: Part II – particulate matter. Atmospheric Environment 42, 6057-6066, 2008.

[6] Appel, K.W., Gilliland, A.B., Sarwar, G., Gilliam, R.C., 2007. Evaluation of the community multiscale air quality (CMAQ) model version 4.5: sensitivities impacting model performance: Part I – ozone. Atmospheric Environment 41, 9603-9615.

[7] Appel, K.W., Roselle, S.J., Gilliam, R.C., and Pleim, J.E.,: Sensitivity of the Community Multiscale Air Quality (CMAQ) model v4.7 results for the eastern United States to MM5 and WRF meteorological drivers. Geoscientific Model Development, 3, 169-188, 2010.

[8] Foley, K.M., Roselle, S.J., Appel, K.W., Bhave, P.V., Pleim, J.E., Otte, T.L., Mathur, R., Sarwar, G., Young, J.O., Gilliam, R.C., Nolte, C.G., Kelly, J.T., Gilliland, A.B., and Bash, J.O.,: Incremental testing of the Community multiscale air quality (CMAQ) modeling system version 4.7. Geoscientific Model Development, 3, 205-226, 2010.

[9] Hogrefe, G., Civeroio, K.L., Hao, W., Ku, J-Y., Zalewsky, E.E., and Sistla, G., Rethinking the Assessment of Photochemical Modeling Systems in Air Quality Planning Applications. Air & Waste Management Assoc., 58:1086-1099, 2008.

[10] Phillips, S., K. Wang, C. Jang, N. Possiel, M. Strum, T. Fox, 2007: Evaluation of 2002 Multi-pollutant Platform: Air Toxics, Ozone, and Particulate Matter, 7th Annual CMAS Conference, Chapel Hill, NC, October 6-8, 2008. (http://www.cmascenter.org/conference/2008/agenda.cfm).

[11] Simon, H., Baker, K.R., and Phillips, S., 2012. Compilation and interpretation of photochemical model performance statistics published between 2006 and 2012. Atmospheric Environment 61, 124-139. http://dx.doi.org/10.1016/j.atmosenv.2012.07.012

[12] Strum, M., Wesson, K., Phillips, S.,Pollack, A., Shepard, S., Jimenez, M., M., Beidler, A., Wilson, M., Ensley, D., Cook, R., Michaels H., and Brzezinski, D. Link Based vs NEI Onroad Emissions Impact on Air Quality Model Predictions. 17th Annual International Emission Inventory Conference, Portland, Oregon, June 2-5, 2008. (http://www.epa.gov/ttn/chief/conference/ei17/session11/strum_pres.pdf)

[13] Tesche, T.W., Morris, R., Tonnesen, G., McNally, D., Boylan, J., Brewer, P., 2006. CMAQ/CAMx annual 2002 performance evaluation over the eastern United States. Atmospheric Environment 40, 4906-4919.

[14] U.S. Environmental Protection Agency; Technical Support Document for the Final Clean Air Interstate Rule: Air Quality Modeling; Office of Air Quality Planning and Standards; RTP, NC; March 2005 (CAIR Docket OAR-2005-0053-2149).

[15] U.S. Environmental Protection Agency, Proposal to Designate an Emissions Control Area for Nitrogen Oxides, Sulfur Oxides, and Particulate Matter: Technical Support Document. EPA-420-R-007, 329pp., 2009. (http://www.epa.gov/otaq/regs/nonroad/marine/ci/420r09007.pdf)

[16] U.S. Environmental Protection Agency, 2010, Renewable Fuel Standard Program (RFS2) Regulatory Impact Analysis. EPA-420-R-10-006. February 2010. Sections 3.4.2.1.2 and 3.4.3.3. Docket EPA-HQ-OAR-2009-0472-11332. (http://www.epa.gov/oms/renewablefuels/420r10006.pdf)

A.2. Evaluation for 8-hour Daily Maximum Ozone

The 8-hour ozone model performance bias and error statistics for each subregion and each season are provided in Table A-2. Spatial plots of the normalized mean bias and error for individual monitors are shown in Figures A-1a through A-1b. The statistics shown in these two figures were calculated over the ozone season using data pairs on days with observed 8-hour ozone of ≥ 60 ppb.

In general, CMAQ slightly under-predicts seasonal eight-hour daily maximum ozone for the five subregions, with the exception of a slight over-prediction in the summer and fall at the Central, Southeast and West subregions (Table A-2). Model performance for 8-hour daily maximum ozone for all subregions is typically better in the spring, summer, and fall months, where the bias statistics are within the range of approximately -0.7 to 12.0 percent and the error statistics range from 12.6 to 23.9 percent The five subregions show relatively similar eight-hour daily maximum ozone performance.

Table A-2. Daily maximum 8-hour ozone performance statistics by subregion, by season for the 2007 CMAQ model simulation.

Subregion	Season	No. of Obs	NMB (%)	NME (%)	FB (%)	FE (%)
Central U.S.	Winter	11,194	-8.1	20.3	-8.4	24.0
	Spring	15,222	-3.2	15.0	-2.4	16.2
	Summer	16,730	11.9	23.9	13.0	25.5
	Fall	14,711	4.5	18.6	5.7	20.0
Midwest	Winter	2,884	-20.9	25.1	-26.0	31.4
	Spring	12,028	-8.3	14.5	-8.4	16.1
	Summer	17,012	-3.3	15.2	-3.4	16.3
	Fall	9,911	-2.0	16.4	0.2	18.1
Southeast	Winter	6,549	-5.3	14.8	-4.0	18.6
	Spring	21,249	-7.0	12.8	-7.1	13.9
	Summer	23,418	3.5	17.1	5.4	18.5
	Fall	17,819	5.9	17.6	7.6	18.9
Northeast	Winter	5,216	-19.7	23.3	-23.6	29.8
	Spring	12,468	-9.2	15.3	-9.5	16.9
	Summer	16,455	-0.7	15.6	-0.4	16.4
	Fall	11,429	0.7	16.9	2.5	18.4
West	Winter	24,485	-0.7	18.2	0.2	20.8
	Spring	28,684	-4.3	12.6	-4.2	13.5
	Summer	32,295	7.1	18.3	6.0	18.5
	Fall	28,984	5.5	17.8	5.9	18.9

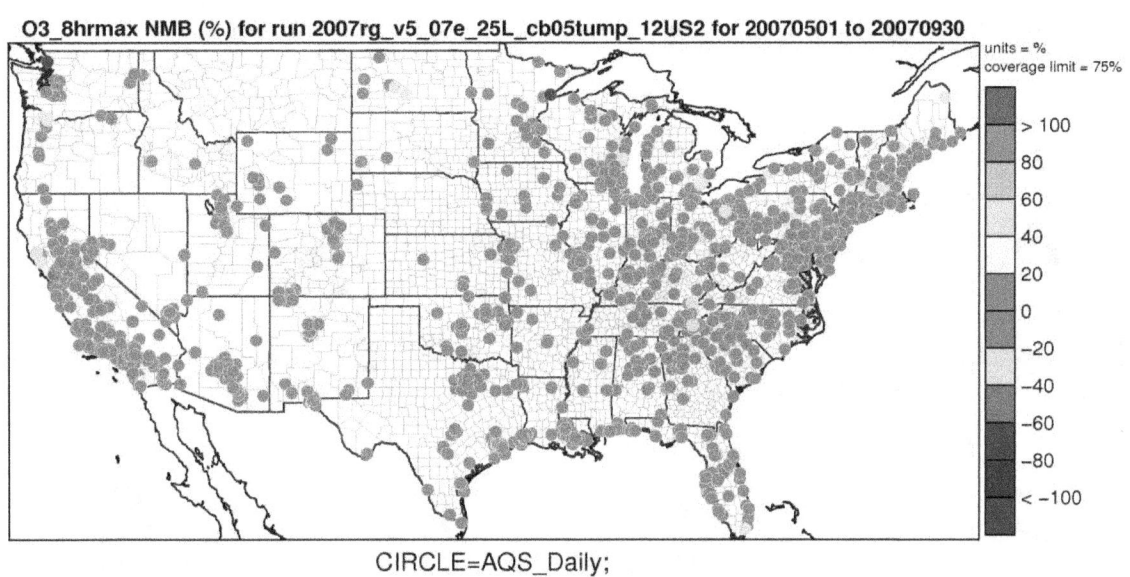

Figure A-1a. Normalized Mean Bias (%) of 8-hour daily maximum ozone greater than 60 ppb over the period May-September 2007 at monitoring sites in the modeling domain.

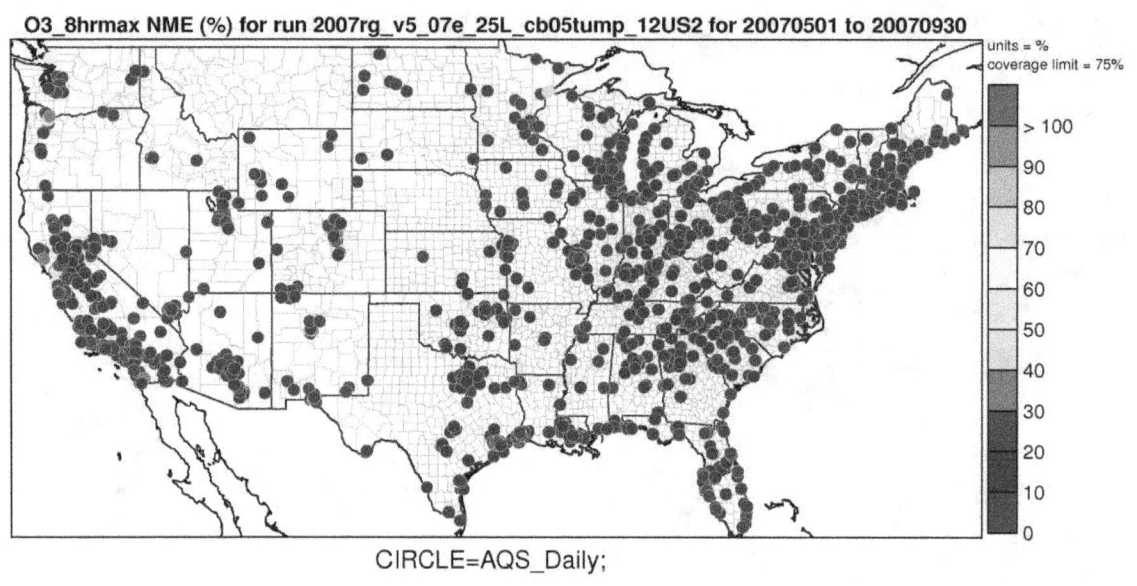

Figure A-1b. Normalized Mean Error (%) of 8-hour daily maximum ozone greater than 60 ppb over the period May-September 2007 at monitoring sites in the modeling domain.

A.3. Evaluation of PM$_{2.5}$ Component Species

The evaluation of 2007 model predictions for PM$_{2.5}$ covers the performance for the individual PM$_{2.5}$ component species (i.e., sulfate, nitrate, organic carbon, elemental carbon, and ammonium). Performance results are provided for each PM$_{2.5}$ species. As indicated above, for each species we present tabular summaries of bias and error statistics by subregion for each season. These statistics are based on the set of observed-predicted pairs of data for the particular quarter at monitoring sites within the subregion. Separate statistics are provided for each monitoring network, as applicable for the particular species measured. For sulfate and nitrate we also provide a more refined temporal and spatial analysis of model performance that includes spatial maps which show the normalized mean bias and error by site, aggregated by season.

A.3.1. Evaluation for Sulfate

The model performance bias and error statistics for sulfate for each subregion and each season are provided in Table A-3. Spatial plots of the normalized mean bias and error by season for individual monitors are shown in Figures A-3 through A-6. As seen in Table A-3, CMAQ generally under-predicts sulfate in the five U.S. subregions throughout the entire year.

Table A-3. Sulfate performance statistics by subregion, by season for the 2007 CMAQ model simulation.

Subregion	Network	Season	No. of Obs.	NMB (%)	NME (%)	FB (%)	FE (%)
Central U.S.	CSN	Winter	771	-15.8	38.3	-14.1	41.7
		Spring	875	-15.2	32.2	-11.3	33.8
		Summer	851	-30.4	42.3	-37.4	54.3
		Fall	587	-10.1	34.9	-3.7	36.8
	IMPROVE	Winter	608	-18.9	40.0	-13.7	43.4
		Spring	722	-17.7	31.4	-11.9	32.4
		Summer	688	-28.2	39.3	-25.8	46.2
		Fall	622	-15.9	31.5	-7.6	37.1
	CASTNet	Winter	72	-32.8	34.3	-34.8	37.4
		Spring	77	-24.6	27.8	-23.6	29.6
		Summer	72	-33.4	37.0	-38.4	46.0
		Fall	75	-21.3	23.8	-19.7	26.4
Midwest	CSN	Winter	598	0.7	38.6	-4.8	38.7
		Spring	637	19.5	43.0	15.3	36.9
		Summer	621	-10.8	28.7	-0.9	30.8
		Fall	639	-12.4	26.7	-4.0	27.5
	IMPROVE	Winter	143	3.5	35.8	-0.1	34.4
		Spring	171	4.7	35.5	6.8	35.2
		Summer	182	-18.8	30.2	-6.2	36.2
		Fall	126	-18.2	27.1	-7.2	31.7
	CASTNet	Winter	142	-13.8	21.8	-16.4	26.6
		Spring	155	-5.9	22.4	-4.4	21.7
		Summer	161	-16.7	22.0	-14.4	24.0
		Fall	157	-20.1	22.7	-16.1	21.8
Southeast	CSN	Winter	888	-4.3	37.1	-3.9	37.0
		Spring	918	-5.3	27.4	-6.1	29.4
		Summer	866	-18.2	32.8	-20.0	39.1
		Fall	911	-10.6	27.8	-6.0	29.5
	IMPROVE	Winter	469	-1.0	36.9	1.1	37.5
		Spring	525	-6.6	29.0	-6.0	31.7
		Summer	500	-24.3	35.7	-31.0	47.1
		Fall	496	-11.9	29.3	-6.3	34.5
	CASTNet	Winter	264	-18.1	22.6	-17.2	23.6
		Spring	292	-13.4	21.3	-14.7	22.9

Subregion	Network	Season	No. of Obs.	NMB (%)	NME (%)	FB (%)	FE (%)
		Summer	268	-21.7	24.9	-28.6	32.9
		Fall	273	-18.6	21.3	-19.3	23.3
Northeast	CSN	Winter	828	-9.1	34.9	-13.0	34.6
		Spring	894	8.2	37.2	4.3	34.9
		Summer	874	-8.9	27.2	-3.1	31.0
		Fall	902	-9.1	28.9	0.0	31.0
	IMPROVE	Winter	561	-6.8	31.1	-10.7	33.2
		Spring	689	7.05	37.9	3.6	38.2
		Summer	649	-13.1	32.3	-4.6	37.7
		Fall	591	-6.7	32.3	7.8	35.5
	CASTNet	Winter	193	-14.5	22.2	-18.6	25.5
		Spring	206	-0.3	25.1	-1.4	26.4
		Summer	192	-15.7	20.6	-12.9	22.1
		Fall	195	-12.3	18.5	-7.2	18.1
West	CSN	Winter	830	-5.5	57.3	1.7	54.3
		Spring	867	-3.8	36.9	0.0	36.1
		Summer	853	-32.3	43.7	-23.5	42.6
		Fall	900	-7.7	47.0	0.3	43.3
	IMPROVE	Winter	2373	22.4	58.3	33.8	56.6
		Spring	2650	-3.6	33.5	3.4	35.2
		Summer	2307	-25.0	41.2	-16.8	42.9
		Fall	2365	-0.6	40.0	11.1	41.2
	CASTNet	Winter	250	6.6	35.9	17.9	37.4
		Spring	273	-18.5	27.1	-17.1	27.7
		Summer	281	-35.3	-36.2	-36.2	41.7
		Fall	268	-10.9	23.6	-5.1	24.3

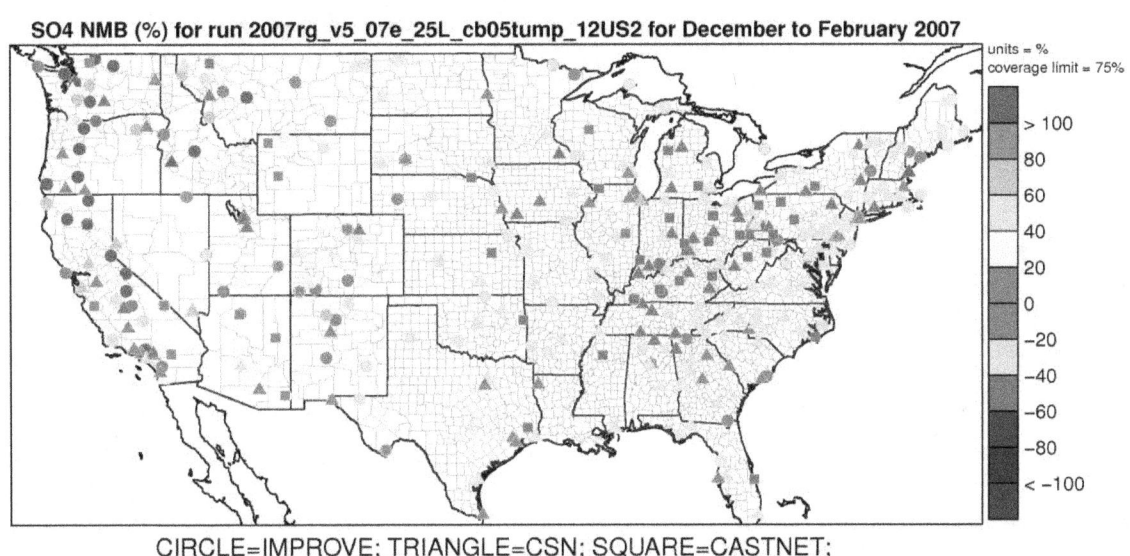

Figure A-3a. Normalized Mean Bias (%) of sulfate during winter 2007 at monitoring sites in the modeling domain.

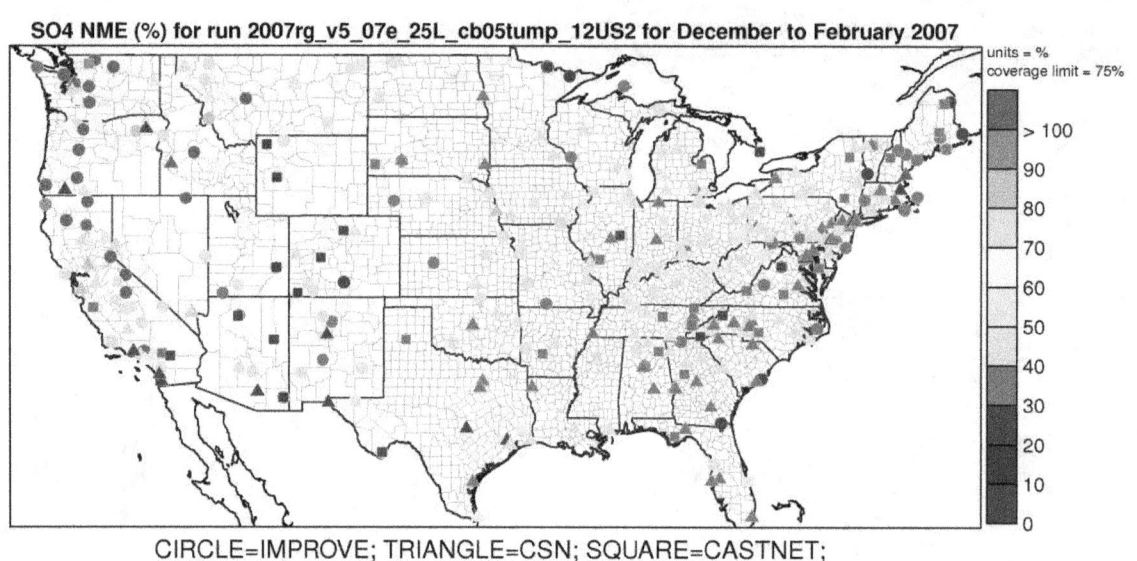

Figure A-3b. Normalized Mean Error (%) of sulfate during winter 2007 at monitoring sites in the modeling domain.

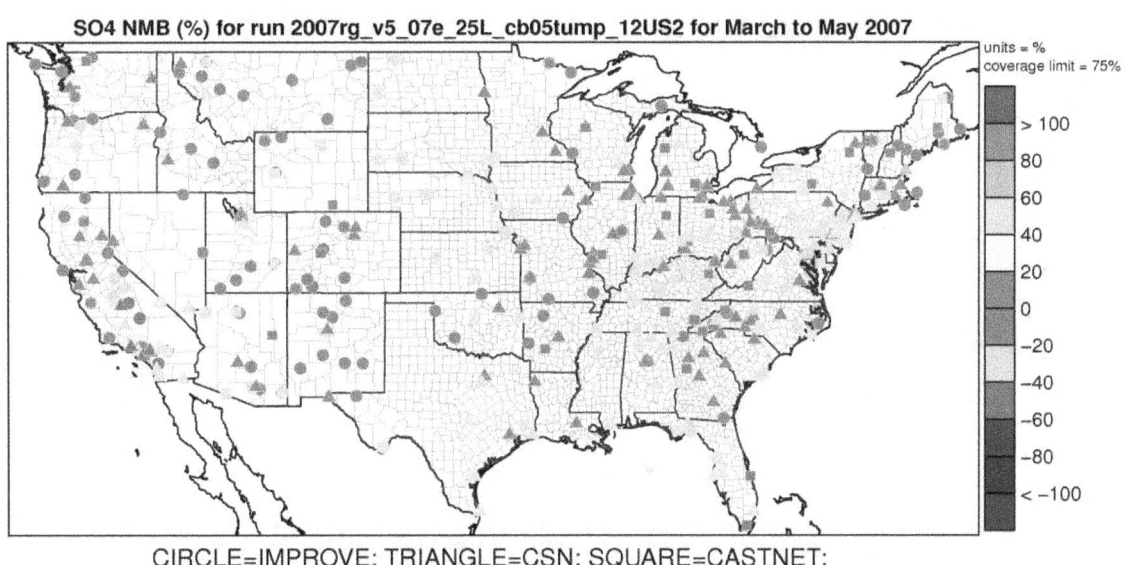

Figure A-4a. Normalized Mean Bias (%) of sulfate during spring 2007 at monitoring sites in the modeling domain.

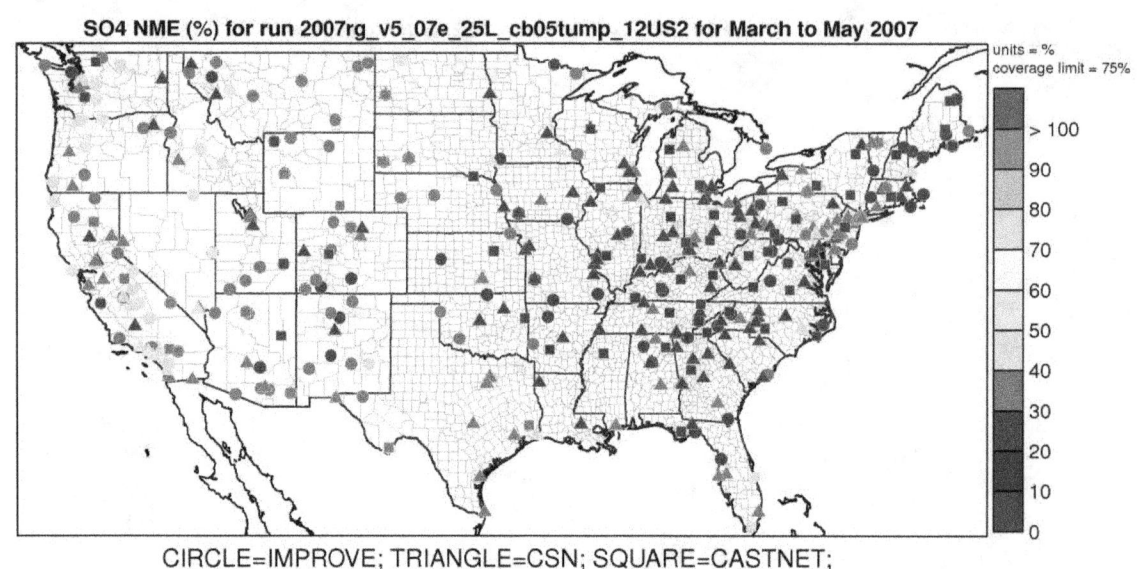

SO4 NME (%) for run 2007rg_v5_07e_25L_cb05tump_12US2 for March to May 2007

units = %
coverage limit = 75%

> 100
90
80
70
60
50
40
30
20
10
0

CIRCLE=IMPROVE; TRIANGLE=CSN; SQUARE=CASTNET;

Figure A-4b. Normalized Mean Error (%) of sulfate during spring 2007 at monitoring sites in the modeling domain.

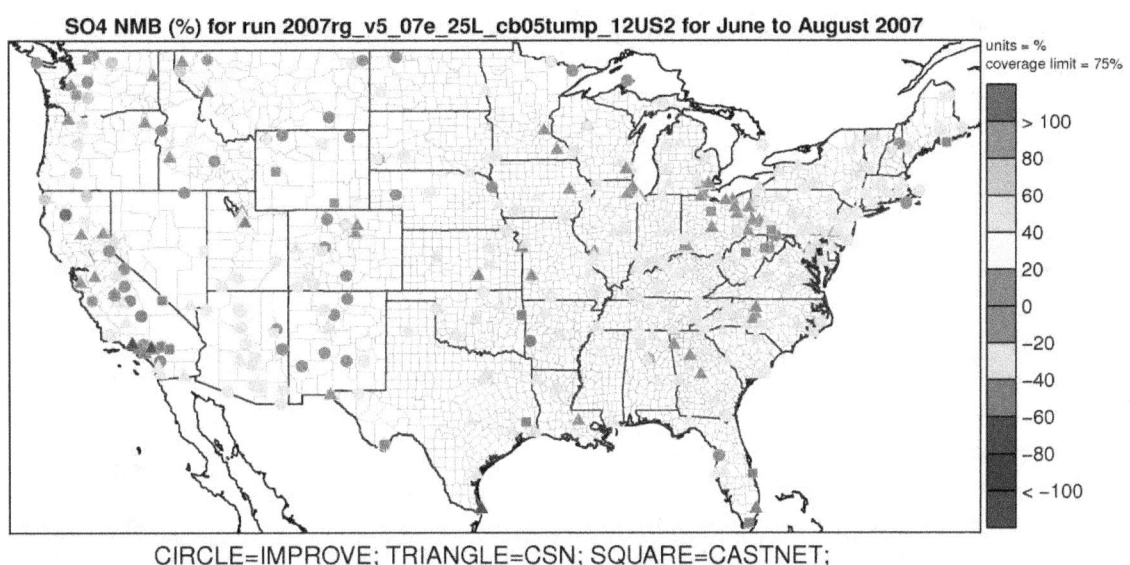

Figure A-5a. Normalized Mean Bias (%) of sulfate during summer 2007 at monitoring sites in the modeling domain.

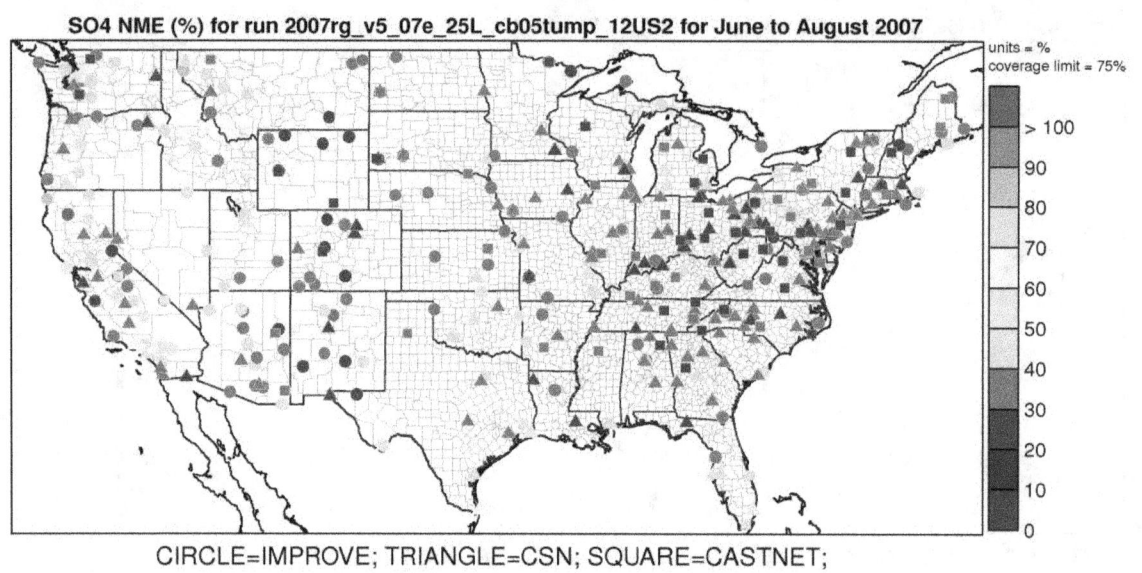

Figure A-5b. Normalized Mean Error (%) of sulfate during summer 2007 at monitoring sites in the modeling domain.

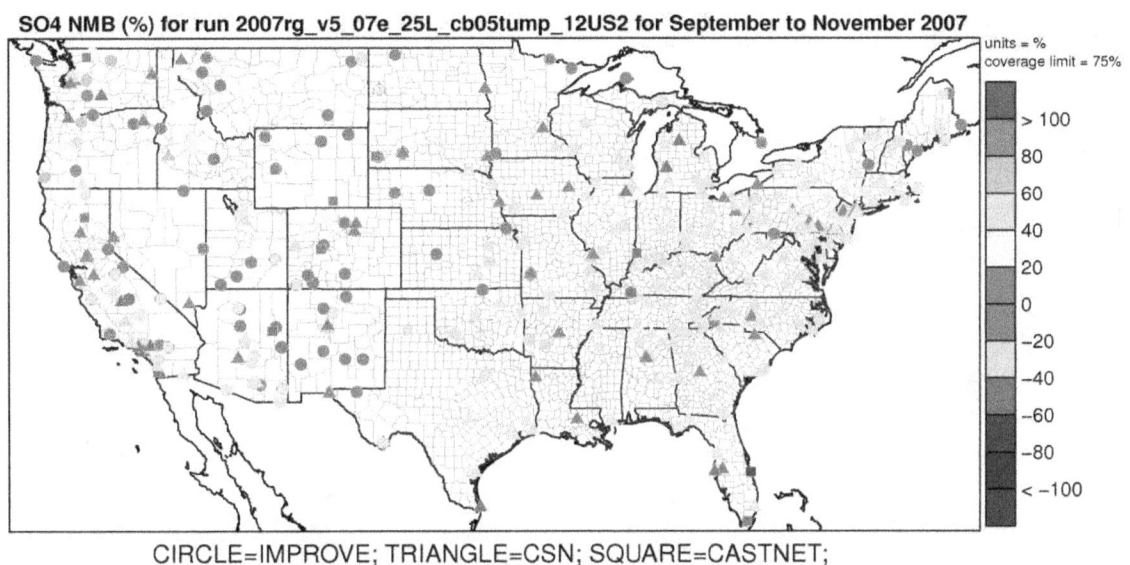

Figure A-6a. Normalized Mean Bias (%) of sulfate during fall 2007 at monitoring sites in the modeling domain.

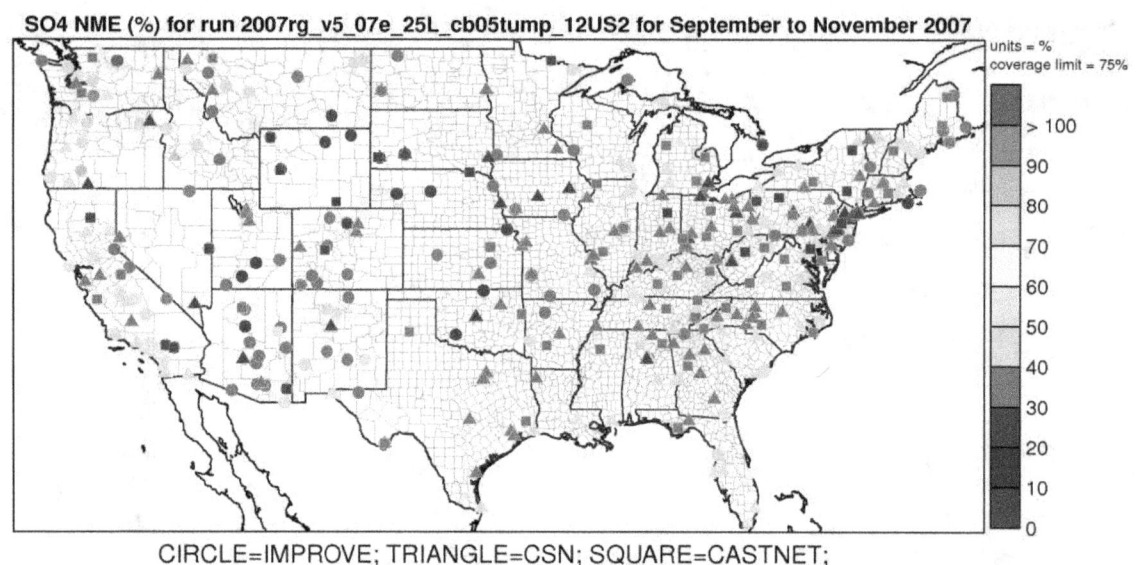

Figure A-6b. Normalized Mean Error (%) of sulfate during fall 2007 at monitoring sites in the modeling domain.

A.3.1. Evaluation for Nitrate

The model performance bias and error statistics for nitrate for each subregion and each season are provided in Table A-4. This table includes statistics for particulate nitrate, as measured at CSN and IMPROVE sites. Spatial plots of the normalized mean bias and error by season for individual monitors are shown in Figures A-7 through A-10. Overall, nitrate performance are over-predicted in the Northeast, Midwest, Southeast and Central U.S.; with the exception at the urban monitors (CSN) where nitrate is under-predicted in the winter. Likewise, nitrate is under-predicted at CSN sites during the summer in the Southeast and Northeast. Model performance shows an under-prediction in the West for all of the seasonal assessments of nitrate.

Table A-4. Nitrate performance statistics by subregion, by season for the 2007 CMAQ model simulation.

Region	Network	Season	No. of Obs.	NMB (%)	NME (%)	FB (%)	FE (%)
Central U.S.	CSN	Winter	479	-7.6	48.7	-9.1	59.8
		Spring	503	26.9	60.3	12.6	65.6
		Summer	485	23.7	99.1	-44.1	95.9
		Fall	460	101.0	129.0	16.0	89.1
	IMPROVE	Winter	608	2.6	54.0	-8.5	70.6
		Spring	722	46.1	76.5	-5.4	90.7
		Summer	688	17.7	109.0	-58.1	112.0
		Fall	622	158.0	188.0	12.4	107.0
Midwest	CSN	Winter	598	-23.7	41.4	-25.3	50.6
		Spring	637	59.1	80.3	38.0	64.6
		Summer	621	38.0	94.3	-13.8	83.3
		Fall	639	64.8	94.9	21.0	74.0
	IMPROVE	Winter	143	-30.1	49.0	-33.0	74.3
		Spring	171	50.4	85.1	-5.8	89.9
		Summer	182	20.3	96.7	-43.8	99.8
		Fall	126	104.0	138.0	-1.5	102.0
Southeast	CSN	Winter	888	-29.3	61.6	-62.9	89.1
		Spring	918	34.4	94.7	-14.6	92.4
		Summer	866	-31.1	83.5	-86.4	115.0
		Fall	911	71.3	136.0	-32.4	109.0
	IMPROVE	Winter	469	-7.3	81.3	-63.8	101.0
		Spring	525	54.9	113.0	-32.1	108.0
		Summer	500	-18.3	109.0	-95.0	136.0
		Fall	496	98.7	179.0	-49.5	126.0
		Fall	273	66.9	76.1	41.7	56.2
Northeast	CSN	Winter	829	-6.4	43.4	-6.6	50.6
		Spring	894	37.5	74.0	28.5	67.5
		Summer	874	-11.2	87.5	-62.7	103.0
		Fall	902	68.5	104.0	-16.2	87.1
	IMPROVE	Winter	561	35.5	74.4	28.5	76.0
		Spring	689	67.2	108.0	28.3	92.4
		Summer	649	5.0	111.0	-64.9	113.0
		Fall	586	108.0	151.0	-12.4	100.0
West	CSN	Winter	831	-47.8	64.8	-65.4	89.7
		Spring	859	-38.9	59.1	-70.9	90.6

Region	Network	Season	No. of Obs.	NMB (%)	NME (%)	FB (%)	FE (%)
		Summer	846	-73.1	76.8	-134.0	138.0
		Fall	896	-49.7	70.7	-69.8	97.5
	IMPROVE	Winter	2,374	-33.1	78.3	-88.0	123.0
		Spring	2,643	-40.3	76.4	-89.9	119.0
		Summer	2,305	-74.6	84.1	-145.0	153.0
		Fall	2,357	-34.2	82.3	-77.2	122.0

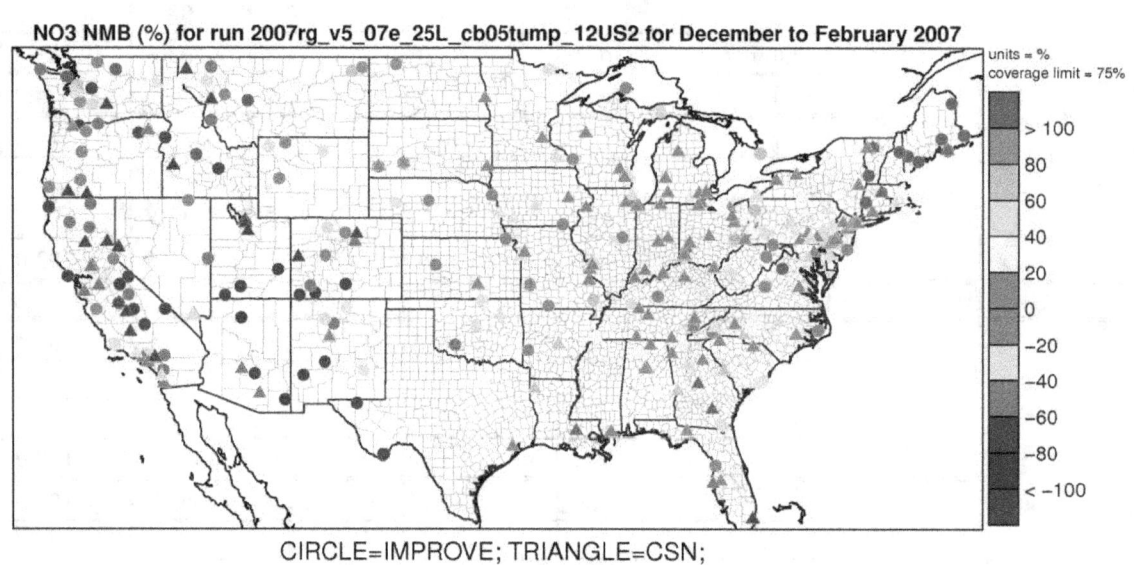

Figure A-7a. Normalized Mean Bias (%) for nitrate during winter 2007 at monitoring sites in the modeling domain.

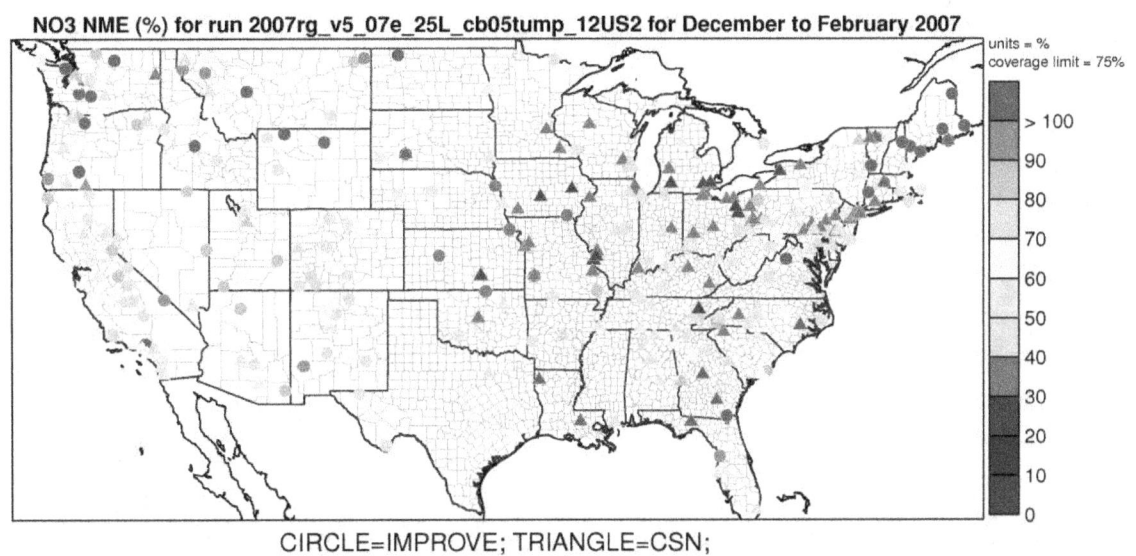

Figure A-7b. Normalized Mean Error (%) for nitrate during winter 2007 at monitoring sites in the modeling domain.

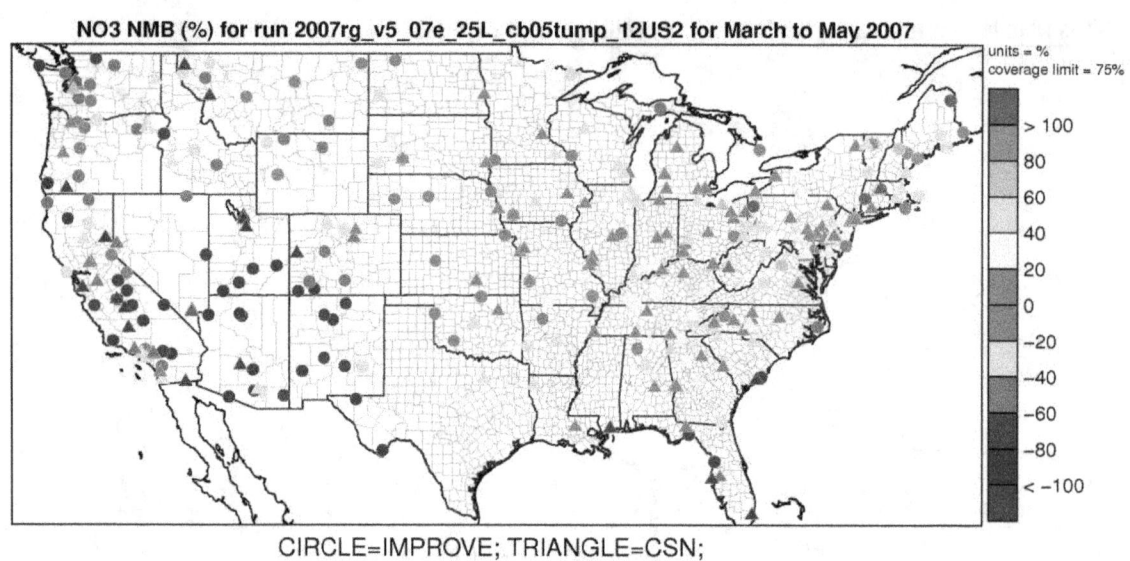

Figure A-8a. Normalized Mean Bias (%) for nitrate during spring 2007 at monitoring sites in the modeling domain.

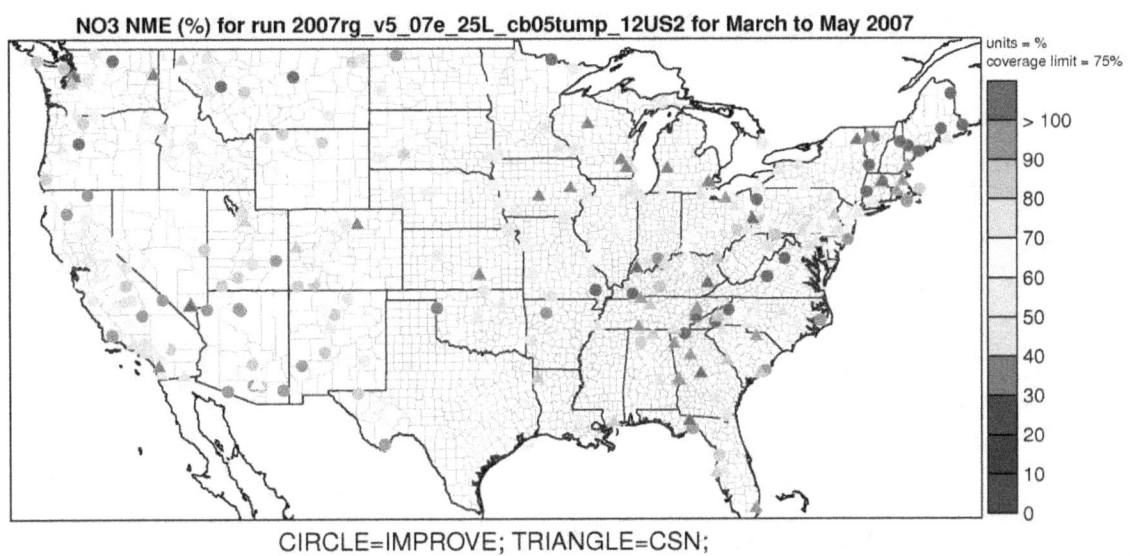

Figure A-8b. Normalized Mean Error (%) for nitrate during spring 2007 at monitoring sites in the modeling domain.

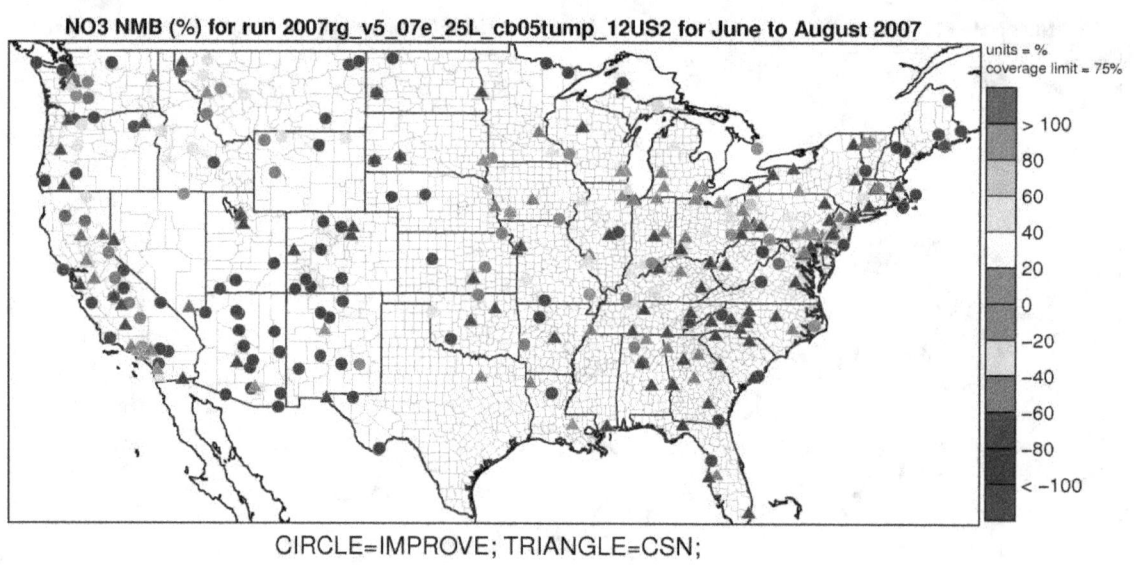

Figure A-9a. Normalized Mean Bias (%) for nitrate during summer 2007 at monitoring sites in the modeling domain.

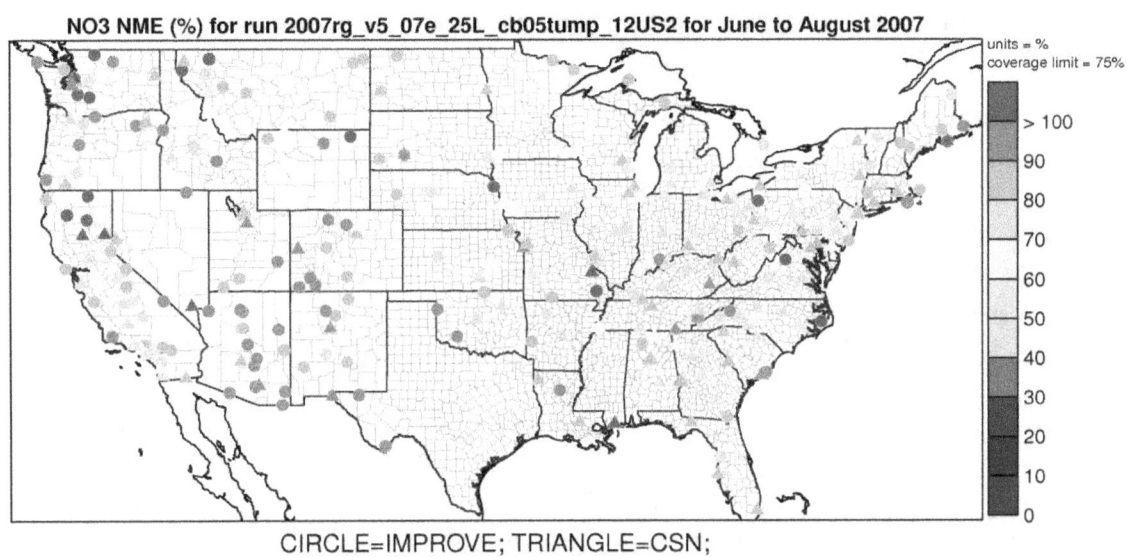

Figure A-9b. Normalized Mean Error (%) for nitrate during summer 2007 at monitoring sites in the modeling domain.

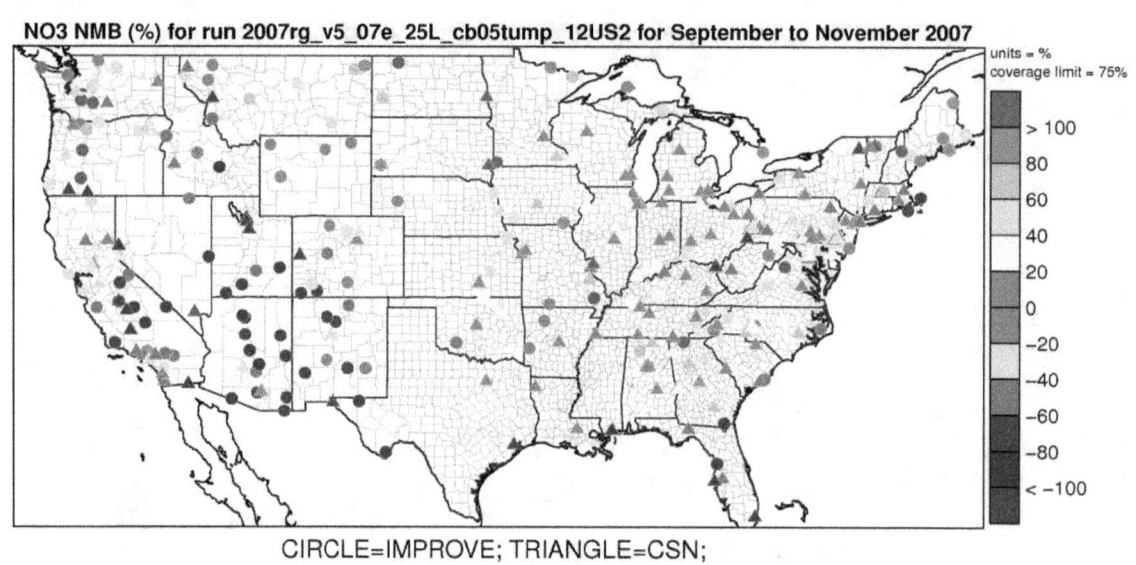

Figure A-10a. **Normalized Mean Bias (%) for nitrate during fall 2007 at monitoring sites in the modeling domain.**

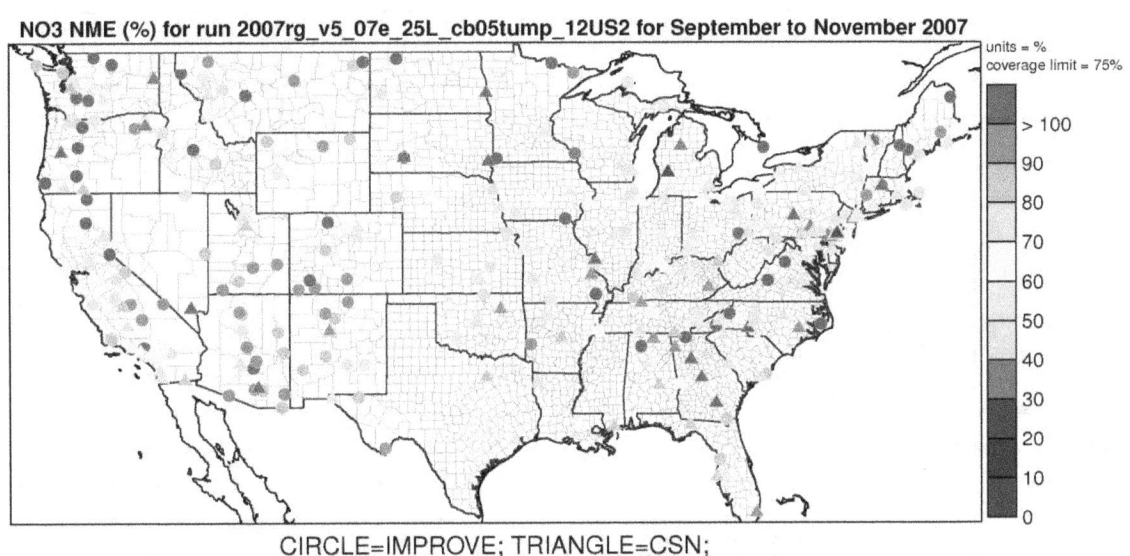

NO3 NME (%) for run 2007rg_v5_07e_25L_cb05tump_12US2 for September to November 2007

CIRCLE=IMPROVE; TRIANGLE=CSN;

Figure A-10b. Normalized Mean Error (%) for nitrate during fall 2007 at monitoring sites in the modeling domain.

H. Seasonal Ammonium Performance

The model performance bias and error statistics for ammonium for each subregion and each season are provided in Table A-5. These statistics indicate model bias for ammonium is generally $\pm$ 40 percent or less for all seasons in each subregion. During the summer, there is slight to moderate under-prediction in the subregions for urban sub-urban locations. In other times of the year ammonium tends to be somewhat over predicted with a bias of 19 percent, on average across the subregions for urban locations.

Table A-5. Ammonium performance statistics by subregion, by season for the 2007 CMAQ model simulation.

Region	Network	Season	No. of Obs.	NMB (%)	NME (%)	FB (%)	FE (%)
Central U.S.	CSN	Winter	771	-2.9	43.3	-1.9	50.7
		Spring	875	4.8	41.9	7.3	43.2
		Summer	851	-21.4	45.9	-24.4	60.9

Region	Network	Season	No. of Obs.	NMB (%)	NME (%)	FB (%)	FE (%)
		Fall	587	17.1	54.8	22.5	55.6
	CASTNet	Winter	72	2.9	37.6	3.7	42.5
		Spring	77	16.6	33.9	10.9	32.2
		Summer	72	-17.1	29.5	-19.8	35.8
		Fall	75	16.9	44.1	24.3	46.3
Midwest	CSN	Winter	598	-10.2	32.2	-5.1	33.9
		Spring	637	47.7	62.2	38.3	50.6
		Summer	621	-0.50	36.9	15.8	41.8
		Fall	639	6.8	37.5	21.2	41.1
	CASTNet	Winter	142	-11.5	24.5	-6.0	25.4
		Spring	155	44.2	51.9	36.5	41.4
		Summer	161	-5.4	25.7	-2.1	27.4
		Fall	157	19.9	45.1	26.7	41.1
Southeast	CSN	Winter	888	-10.9	41.2	-11.0	44.5
		Spring	918	8.0	39.4	7.9	40.2
		Summer	866	-14.4	36.8	-9.1	44.4
		Fall	911	2.5	42.2	13.1	45.5
	CASTNet	Winter	264	-7.1	28.0	-7.6	29.7
		Spring	292	8.2	30.9	6.6	30.7
		Summer	268	-32.0	35.4	-45.2	48.8
		Fall	273	-9.0	36.4	-7.5	41.0
Northeast	CSN	Winter	828	0.1	34.1	4.2	34.3
		Spring	894	31.1	53.2	34.0	49.5
		Summer	874	-11.5	36.1	3.6	44.0
		Fall	902	16.6	49.4	28.4	50.6
	CASTNet	Winter	193	21.3	37.6	25.9	36.8
		Spring	206	42.0	48.5	32.0	38.3
		Summer	192	-23.5	29.8	-26.7	34.7
		Fall	195	8.7	39.0	13.6	36.2
West	CSN	Winter	829	-30.8	60.8	-15.1	65.9
		Spring	859	-1.5	52.6	17.8	51.2
		Summer	849	-33.3	53.1	-5.1	51.7
		Fall	886	-22.9	63.6	8.1	58.4
	CASTNet	Winter	250	-4.0	40.8	6.2	39.3
		Spring	273	-9.6	32.0	-5.2	31.7
		Summer	281	-33.7	40.5	-34.9	44.9
		Fall	268	-4.1	31.8	0.9	31.2

I. Seasonal Elemental Carbon Performance

The model performance bias and error statistics for elemental carbon for each subregion and each season are provided in Table A-6. The statistics show clear over prediction at urban sites in all subregions. For example, NMBs typically range between 50 and 100 percent at urban sites in the Midwest, Northeast, and Central subregions with only slightly less over prediction at urban sites in the Southeast. Rural sites show much less over prediction than at urban sites with under predictions occurring in the spring, summer, and fall at rural sites in the Southeast, Midwest and Central subregions. In the West, the model tends to over predict at both urban and rural sites during all seasons. In addition, the predictions for urban sites have greater error than the predictions for rural locations in the West.

Table A-6. Elemental Carbon performance statistics by subregion, by season for the 2007 CMAQ model simulation.

Subregion	Network	Season	No. of Obs.	NMB (%)	NME (%)	FB (%)	FE (%)
Central U.S.	CSN	Winter	816	103.0	136.0	56.8	78.1
		Spring	938	94.0	117.0	46.3	71.2
		Summer	875	113.0	136.0	43.0	81.2
		Fall	618	96.8	115.0	58.0	71.8
	IMPROVE	Winter	589	9.4	54.5	4.4	47.1
		Spring	716	-9.0	56.0	-9.9	53.8
		Summer	701	-30.3	46.8	-38.2	56.2
		Fall	620	-17.1	34.8	-16.0	41.1
Midwest	CSN	Winter	602	121.0	136.0	68.6	76.0
		Spring	637	65.0	86.1	49.2	61.8
		Summer	621	49.3	65.7	38.7	54.8
		Fall	642	53.8	73.8	40.1	55.9
	IMPROVE	Winter	182	61.6	80.0	22.6	45.9
		Spring	184	19.0	57.8	-11.4	51.3
		Summer	185	-13.1	41.3	-36.9	53.9
		Fall	145	-12.7	33.6	-19.2	48.2
Southeast	CSN	Winter	889	38.5	62.4	30.7	49.6
		Spring	914	38.7	63.7	37.4	54.6
		Summer	866	41.4	69.8	38.4	61.4
		Fall	909	13.3	46.4	19.1	46.0
	IMPROVE	Winter	491	-3.0	44.5	-1.0	48.7
		Spring	530	-16.5	44.9	-11.0	45.1
		Summer	493	-40.9	48.2	-55.5	71.5
		Fall	481	-26.5	38.8	-22.5	45.5
Northeast	CSN	Winter	831	98.5	111.0	57.6	67.0

Subregion	Network	Season	No. of Obs.	NMB (%)	NME (%)	FB (%)	FE (%)
		Spring	881	92.6	109.0	57.8	69.3
		Summer	866	66.9	89.6	46.2	63.8
		Fall	901	54.3	84.2	35.6	57.1
	IMPROVE	Winter	603	46.1	73.8	22.3	53.1
		Spring	658	29.2	64.0	11.7	54.6
		Summer	596	-19.7	45.8	-37.2	57.3
		Fall	591	32.9	59.1	6.7	49.7
West	CSN	Winter	808	50.2	89.1	24.3	67.6
		Spring	822	111.0	134.0	47.8	76.7
		Summer	806	121.0	134.0	60.3	74.4
		Fall	867	58.8	91.4	29.6	65.9
	IMPROVE	Winter	2,338	1.8	65.1	-15.8	64.8
		Spring	2,597	19.4	69.7	-1.5	54.2
		Summer	2,314	30.0	77.9	18.4	58.6
		Fall	2,372	9.0	67.4	-9.5	59.6

J. Seasonal Organic Carbon Performance

The model performance bias and error statistics for organic carbon for each subregion and each season are provided in Table A-7. The statistics in this table indicate a tendency for the modeling platform to somewhat under predict observed organic carbon concentrations during the spring, summer, and fall at urban and rural locations across the Eastern subregions. Likewise, the modeling platform under predicts organic carbon during all seasons at urban and rural locations in the Western subregion, except in the summer at rural sites. These biases and errors reflect sampling artifacts among each monitoring network. In addition, uncertainties exist for primary organic mass emissions and secondary organic aerosol formation. Research efforts are ongoing to improve fire emission estimates and understand the formation of semi-volatile compounds, and the partitioning of SOA between the gas and particulate phases.

Table A-7. Organic Carbon performance statistics by subregion, by season for the 2007 CMAQ model simulation.

Region	Network	Season	No. of Obs.	NMB (%)	NME (%)	FB (%)	FE (%)
Central U.S.	CSN	Winter	544	-2.0	57.1	12.9	59.6
		Spring	628	-35.3	52.6	-32.8	63.7
		Summer	595	-51.9	54.5	-70.7	77.1
		Fall	493	-31.7	45.6	-29.2	57.2
	IMPROVE	Winter	589	-9.0	51.2	-13.0	48.1
		Spring	715	-38.7	57.7	-38.4	61.3
		Summer	699	-50.3	52.6	-70.3	74.6
		Fall	619	-44.7	48.4	-54.8	62.7
Midwest	CSN	Winter	566	1.1	52.3	19.1	53.5
		Spring	605	-29.4	45.9	-17.8	52.8
		Summer	619	-53.8	55.1	-70.8	74.2
		Fall	595	-29.7	41.7	-17.9	52.5
	IMPROVE	Winter	182	0.9	37.7	0.0	37.2
		Spring	184	-25.9	36.4	-32.9	44.6
		Summer	185	-49.0	52.0	-65.7	69.8
		Fall	144	-35.6	44.0	-44.5	62.2
Southeast	CSN	Winter	871	-26.8	45.7	-16.5	51.0
		Spring	901	-36.0	48.9	-29.4	57.3
		Summer	857	-56.2	58.1	-76.7	81.4
		Fall	880	-40.5	46.4	-43.7	57.9
	IMPROVE	Winter	491	-11.0	45.1	-12.5	51.2
		Spring	529	-9.6	49.2	-15.6	50.5
		Summer	492	-49.0	54.5	-67.2	75.6
		Fall	481	-34.4	41.5	-42.3	53.6
Northeast	CSN	Winter	806	25.8	58.4	29.7	54.8
		Spring	832	1.9	50.8	8.1	53.1

Region	Network	Season	No. of Obs.	NMB (%)	NME (%)	FB (%)	FE (%)
		Summer	859	-47.4	51.8	-61.4	69.5
		Fall	830	-4.9	47.3	3.2	53.3
	IMPROVE	Winter	602	46.4	68.1	30.6	51.7
		Spring	657	3.1	46.1	-3.6	46.1
		Summer	596	-47.2	51.6	-59.7	66.6
		Fall	588	13.9	47.4	-2.3	44.0
	CSN	Winter	803	25.2	67.4	-19.3	70.0
		Spring	823	-9.2	60.3	-1.0	60.3
West		Summer	840	-22.3	41.3	-26.4	49.9
		Fall	881	-26.5	56.5	-24.2	58.0
	IMPROVE	Winter	2,296	-17.0	58.9	-23.2	64.7
		Spring	2,559	-22.6	51.5	-24.8	56.6
		Summer	2,297	4.7	65.2	-0.9	60.1
		Fall	2,350	-21.4	56.8	-26.5	62.1

K. Seasonal Hazardous Air Pollutants Performance

A seasonal operational model performance evaluation for specific hazardous air pollutants (formaldehyde, acetaldehyde, benzene, acrolein, and 1,3-butadiene) was conducted in order to estimate the ability of the CMAQ modeling system to replicate the base year concentrations for the 12 km Continental United States domain. The seasonal model performance results for the East and West are presented below in Tables A-8 and A-9, respectively. Toxic measurements included in the evaluation were taken from the 2007 State/local monitoring site data in the National Air Toxics Trends Stations (NATTS). Similar to $PM_{2.5}$ and ozone, the evaluation principally consists of statistical assessments of model versus observed pairs that were paired in time and space on daily basis.

Model predictions of annual formaldehyde, acetaldehyde and benzene showed relatively small to moderate bias and error percentages when compared to observations. The model yielded larger bias and error results for 1,3 butadiene and acrolein based on limited monitoring sites. Model performance for HAPs is not as good as model performance for ozone and $PM_{2.5}$. Technical issues in the HAPs data consist of (1) uncertainties in monitoring methods; (2) limited measurements in time/space to characterize ambient concentrations ("local in nature"); (3) commensurability issues between measurements and model predictions; (4) emissions and science uncertainty issues may also affect model performance; and (5) limited data for estimating intercontinental transport that effects the estimation of boundary conditions (i.e., boundary estimates for some species are much higher than predicted values inside the domain).

As with the national, annual $PM_{2.5}$ and ozone CMAQ modeling, the "acceptability" of model performance was judged by comparing our CMAQ 2007 performance results to the limited performance found in recent regional multi-pollutant model applications.[17,18,19] Overall, the normalized mean bias and error (NMB and NME), as well as the fractional bias and error (FB and FE) statistics shown below indicate that CMAQ-predicted 2007 toxics (i.e., observation vs. model predictions) are within the range of recent regional modeling applications.

---

[17] Phillips, S., K. Wang, C. Jang, N. Possiel, M. Strum, T. Fox, 2007: Evaluation of 2002 Multi-pollutant Platform: Air Toxics, Ozone, and Particulate Matter, 7th Annual CMAS Conference, Chapel Hill, NC, October 6-8, 2008.

[18] Strum, M., Wesson, K., Phillips, S., Cook, R., Michaels, H., Brzezinski, D., Pollack, A., Jimenez, M., Shepard, S. Impact of using lin-level emissions on multi-pollutant air quality model predictions at regional and local scales. 17th Annual International Emission Inventory Conference, Portland, Oregon , June 2-5, 2008.

[19] Wesson, K., N. Fann, and B. Timin, 2010: Draft Manuscript: Air Quality and Benefits Model Responsiveness to Varying Horizontal Resolution in the Detroit Urban Area, Atmospheric Pollution Research, Special Issue: Air Quality Modeling and Analysis.

Table A-8. Air toxics performance statistics by season for the 2007 CMAQ model simulation.

Air Toxic Species	Season	No. of Obs.	NMB (%)	NME (%)	FB (%)	FE (%)
Formaldehyde	Winter	613	-48.6	60.5	-53.0	68.0
	Spring	435	-51.0	62.0	-51.5	74.6
	Summer	745	-51.6	65.4	-31.0	57.3
	Fall	622	-51.1	62.0	-39.0	60.4
Acetaldehyde	Winter	577	-31.0	48.2	-31.0	54.0
	Spring	387	-14.6	45.2	-13.9	49.2
	Summer	421	-32.0	53.3	-28.1	58.0
	Fall	455	-22.8	57.3	-21.8	55.4
Benzene	Winter	1,507	8.4	65.2	12.5	53.1
	Spring	1,122	1.4	63.1	3.4	54.6
	Summer	1,038	15.5	68.2	12.3	66.2
	Fall	938	25.6	64.7	19.2	59.7
1,3-Butadiene	Winter	1,385	-30.4	91.2	9.5	85.1
	Spring	1,033	-41.6	88.4	-17.6	74.4
	Summer	1,522	-51.2	88.9	-47.3	89.4
	Fall	1,257	-33.4	81.8	-21.2	87.6
Acrolein	Winter	559	-91.5	93.3	-156.0	156.0
	Spring	416	-93.6	94.9	-169.0	169.0
	Summer	685	-94.1	99.0	-155.0	155.0
	Fall	951	-95.2	98.8	-150.0	154.0

L. Seasonal Nitrate and Sulfate Deposition Performance

Seasonal nitrate and sulfate deposition performance statistics for the 12 km Continental U.S. domain are provided in Table A-10. The model predictions for seasonal nitrate deposition generally show under-predictions for the continental U.S. NADP sites (NMB values range from -6% to -34%). Sulfate deposition performance shows the similar predictions (NMB values range from -12% to 28%). The errors for both annual nitrate and sulfate are relatively moderate with values ranging from 51% to 70% which reflect scatter in the model predictions versus observation comparison.

Table A-10. Nitrate and sulfate wet deposition performance statistics by season for the 2007 CMAQ model simulation.

Wet Deposition Species	Season	No. of Obs.	NMB (%)	NME (%)	FB (%)	FE (%)
Nitrate	Winter	1,992	-5.9	60.7	-21.0	75.5
	Spring	2,013	-21.6	51.3	-27.4	70.5
	Summer	2,147	-34.3	60.9	-34.6	77.4
	Fall	2,037	-14.8	57.0	-25.0	74.8
Sulfate	Winter	1,992	-27.5	53.5	-30.1	76.2
	Spring	2,013	-17.6	53.1	-13.0	70.7
	Summer	2,147	-11.5	69.8	-5.7	78.7
	Fall	2,037	-21.1	59.2	-28.2	78.0

Air Quality Modeling Technical Support Document: Tier 3 Motor Vehicle Emission and Fuel Standards

Appendix B

8-Hour Ozone Design Values for Air Quality Modeling Scenarios

U.S. Environmental Protection Agency
Office of Air Quality Planning and Standards
Air Quality Assessment Division
Research Triangle Park, NC 27711
February 2014

Table B-1. 8-Hour Ozone Design Values for Tier3 Scenarios
(units are ppb)

State	County	2007 Baseline DV	2018 Reference DV	2018 Tier 3 Control DV	2030 Reference DV	2030 Tier 3 Control DV
Alabama	Baldwin	75.7	64.05	63.67	59.05	58.25
Alabama	Clay	76.0	60.88	60.16	57.29	56.19
Alabama	Colbert	71.0	55.99	55.43	52.73	51.82
Alabama	Elmore	69.7	55.92	55.18	52.26	51.14
Alabama	Etowah	70.0	55.18	54.50	51.73	50.65
Alabama	Houston	68.7	56.30	55.68	52.97	52.01
Alabama	Jefferson	85.3	69.27	68.61	65.75	64.7
Alabama	Lawrence	74.0	60.85	60.30	57.51	56.61
Alabama	Madison	76.3	60.91	60.21	56.84	55.71
Alabama	Mobile	76.7	66.13	65.76	61.42	60.73
Alabama	Montgomery	73.0	58.63	57.86	54.74	53.58
Alabama	Morgan	74.7	60.70	60.09	57.07	56.1
Alabama	Russell	73.0	57.31	56.45	52.76	51.56
Alabama	Shelby	85.3	67.04	66.07	62.29	60.66
Alabama	Sumter	64.0	53.97	53.52	51.21	50.48
Alabama	Tuscaloosa	73.0	59.65	59.06	56.34	55.41
Arizona	Cochise	68.7	60.43	60.33	58.86	58.37
Arizona	Coconino	70.0	64.28	64.22	62.49	62.4
Arizona	Gila	77.7	63.80	63.72	60.27	59.24
Arizona	La Paz	72.5	63.00	62.96	60.38	60.24
Arizona	Maricopa	79.7	67.47	67.36	64.13	62.86
Arizona	Pima	73.7	61.33	60.64	57.1	55.51
Arizona	Pinal	77.3	64.24	64.14	60.8	59.75
Arizona	Yuma	74.5	62.98	62.66	59.8	59.21
Arkansas	Crittenden	82.3	69.40	68.75	65.62	64.35
Arkansas	Newton	70.3	59.83	59.39	57.24	56.51
Arkansas	Polk	73.3	62.34	61.90	59.55	58.82
Arkansas	Pulaski	78.7	64.53	63.81	60.54	59.3
Arkansas	Washington	64.0	52.03	51.34	48.68	47.39
California	Alameda	78.7	67.38	67.35	62.13	62.1
California	Amador	82.3	67.30	67.28	60.25	60.22
California	Butte	83.7	69.01	68.99	62.11	62.08
California	Calaveras	87.0	71.89	71.87	64.94	64.91
California	Colusa	68.0	57.33	57.30	52.2	52.16
California	Contra Costa	75.0	68.97	68.94	64.94	64.89
California	El Dorado	95.7	77.00	76.97	68.38	68.35

State	County	2007 Baseline DV	2018 Reference DV	2018 Tier 3 Control DV	2030 Reference DV	2030 Tier 3 Control DV
California	Fresno	99.7	83.66	83.65	76.1	76.09
California	Glenn	69.0	58.42	58.38	53.5	53.46
California	Imperial	82.7	69.74	69.72	70.72	70.68
California	Inyo	80.7	69.47	69.45	65.61	65.56
California	Kern	106.7	88.37	88.35	82.43	82.41
California	Kings	90.5	74.67	74.65	67.48	67.46
California	Lake	61.3	51.25	51.22	47.37	47.34
California	Los Angeles	105.7	100.93	100.91	93.96	93.86
California	Madera	81.7	69.85	69.83	63.38	63.36
California	Mariposa	86.3	71.74	71.72	66.33	66.32
California	Merced	89.3	73.89	73.86	66.58	66.54
California	Monterey	58.3	51.16	51.15	48.44	48.42
California	Napa	59.7	50.41	50.37	45.95	45.91
California	Nevada	91.0	73.80	73.77	65.99	65.97
California	Orange	85.0	76.10	76.08	72.75	72.65
California	Placer	89.3	71.70	71.69	63.11	63.09
California	Riverside	104.5	95.72	95.70	90.46	90.36
California	Sacramento	100.0	80.17	80.15	70.46	70.44
California	San Benito	76.7	63.86	63.83	58.99	58.96
California	San Bernardino	119.7	109.19	109.16	102.19	102.09
California	San Diego	90.0	73.83	73.81	67.09	67.05
California	San Joaquin	85.0	70.06	70.04	63.78	63.76
California	San Luis Obispo	84.0	69.43	69.41	64.01	63.99
California	Santa Barbara	74.0	65.46	65.45	60.81	60.79
California	Santa Clara	74.3	61.34	61.32	57.04	57
California	Santa Cruz	60.0	53.61	53.59	50.54	50.5
California	Shasta	76.0	64.22	64.18	59.34	59.31
California	Solano	73.7	61.76	61.73	56.26	56.22
California	Sonoma	57.0	43.02	42.98	37.53	37.5
California	Stanislaus	87.3	73.48	73.44	66.87	66.83
California	Sutter	81.7	69.52	69.49	63.44	63.4
California	Tehama	83.3	69.22	69.17	63.21	63.16
California	Tulare	103.7	83.33	83.32	75.19	75.17
California	Tuolumne	84.7	71.18	71.16	65.26	65.24
California	Ventura	87.7	76.33	76.31	68.94	68.91
California	Yolo	77.7	66.06	66.02	59.87	59.84
Colorado	Adams	71.0	63.19	62.57	60.73	58.96
Colorado	Arapahoe	78.0	68.09	67.32	65.26	63.39

State	County	2007 Baseline DV	2018 Reference DV	2018 Tier 3 Control DV	2030 Reference DV	2030 Tier 3 Control DV
Colorado	Boulder	80.0	69.40	68.69	66.41	64.96
Colorado	Denver	72.3	64.35	63.72	61.84	60.04
Colorado	Douglas	81.0	70.95	70.13	67.88	65.82
Colorado	El Paso	72.0	64.64	64.31	63	62.37
Colorado	Jefferson	84.3	74.76	73.99	71.54	69.39
Colorado	La Plata	70.3	62.65	62.55	61.88	61.72
Colorado	Larimer	80.0	70.89	70.48	68.54	67.7
Colorado	Montezuma	71.0	66.97	66.86	66.28	66.07
Colorado	Weld	75.0	68.28	67.98	66.6	65.96
Connecticut	Fairfield	88.7	76.03	75.16	72.92	72.36
Connecticut	Hartford	86.0	71.01	70.03	65.28	64.28
Connecticut	Litchfield	83.3	67.35	66.42	62.11	61.13
Connecticut	Middlesex	87.0	74.68	73.91	69.72	68.81
Connecticut	New Haven	87.3	72.78	72.28	68.77	68.09
Connecticut	New London	88.0	70.21	69.82	66.23	65.68
Connecticut	Tolland	86.0	70.82	69.90	65.35	64.37
Delaware	Kent	79.0	66.08	65.57	60.41	59.7
Delaware	New Castle	81.3	68.57	67.85	63.84	62.87
Delaware	Sussex	79.7	65.40	64.70	61.05	60.04
D.C.	Washington	84.7	69.21	68.03	65.47	62.74
Florida	Alachua	70.5	54.26	53.49	50.23	48.9
Florida	Baker	66.7	53.64	53.05	50.14	49.17
Florida	Bay	75.0	63.09	62.64	59.54	58.79
Florida	Brevard	69.7	57.22	56.65	53.29	52.24
Florida	Broward	66.0	59.30	58.91	56.25	54.78
Florida	Collier	69.0	56.06	55.21	51.54	49.94
Florida	Columbia	69.0	55.59	54.95	51.81	50.78
Florida	Duval	74.0	59.50	58.76	54.7	53.36
Florida	Escambia	78.7	64.63	63.95	59.69	58.42
Florida	Highlands	71.0	61.73	61.32	58.85	58.16
Florida	Hillsborough	80.0	69.69	69.01	65.11	63.75
Florida	Holmes	69.3	57.29	56.72	53.96	53.09
Florida	Lake	73.7	61.82	61.04	58.23	56.69
Florida	Lee	67.7	55.23	54.53	51.05	49.77
Florida	Leon	70.3	54.96	54.04	50.31	48.76
Florida	Manatee	77.0	63.48	62.60	57.6	55.67
Florida	Marion	70.7	55.69	54.95	52.03	50.93
Florida	Miami-Dade	72.0	64.38	63.95	60.81	59.42

State	County	2007 Baseline DV	2018 Reference DV	2018 Tier 3 Control DV	2030 Reference DV	2030 Tier 3 Control DV
Florida	Orange	75.7	64.29	63.53	60.86	59.28
Florida	Osceola	71.3	59.67	58.86	55.93	54.15
Florida	Palm Beach	66.0	58.15	57.65	54.68	53.44
Florida	Pasco	74.7	61.97	61.29	58.49	57.32
Florida	Pinellas	70.3	59.77	59.22	56.16	55.05
Florida	Polk	73.7	58.82	57.95	54.21	52.59
Florida	St Lucie	64.3	55.29	54.83	52.09	51.15
Florida	Santa Rosa	79.3	64.00	63.33	58.85	57.6
Florida	Sarasota	75.3	60.98	60.11	54.84	53.02
Florida	Seminole	71.0	59.50	58.67	55.52	53.83
Florida	Volusia	65.7	52.50	51.75	48.62	47.44
Florida	Wakulla	69.7	59.27	58.84	56.19	55.46
Georgia	Bibb	80.5	62.14	61.15	57.09	55.71
Georgia	Chatham	66.0	54.24	53.65	49.62	48.67
Georgia	Chattooga	73.3	57.06	56.22	52.07	50.66
Georgia	Clarke	80.0	59.18	58.00	53.2	51.29
Georgia	Cobb	84.0	65.75	64.47	59.61	57.26
Georgia	Columbia	72.7	58.86	58.02	54	52.65
Georgia	Coweta	82.0	65.13	64.19	61.23	59.82
Georgia	Dawson	76.3	58.89	57.79	53.14	51.22
Georgia	De Kalb	90.7	71.78	70.35	65.29	62.55
Georgia	Douglas	85.3	65.73	64.46	59.1	57.02
Georgia	Fayette	87.5	67.51	66.12	61.53	59.08
Georgia	Fulton	90.3	71.01	69.52	64.24	61.51
Georgia	Glynn	64.3	51.27	50.64	46.88	45.86
Georgia	Gwinnett	86.0	65.06	63.55	57.43	54.61
Georgia	Henry	92.0	69.43	67.87	62.76	60.22
Georgia	Murray	77.7	58.67	57.68	52.46	50.91
Georgia	Muscogee	77.0	60.42	59.45	55.16	53.8
Georgia	Paulding	79.0	60.79	59.73	55.18	53.35
Georgia	Richmond	77.7	62.99	62.09	57.75	56.37
Georgia	Rockdale	91.7	69.15	67.62	62.32	59.79
Georgia	Sumter	71.7	57.94	57.25	53.87	52.87
Idaho	Ada	77.0	65.56	64.41	60.96	58.7
Idaho	Butte	64.0	60.10	59.98	58.73	58.53
Idaho	Kootenai	63.5	54.18	53.61	48.2	47.4
Illinois	Adams	67.0	56.68	56.24	53.56	52.76
Illinois	Champaign	66.3	57.44	57.06	55.03	54.33

State	County	2007 Baseline DV	2018 Reference DV	2018 Tier 3 Control DV	2030 Reference DV	2030 Tier 3 Control DV
Illinois	Clark	66.0	57.00	56.62	54.56	53.88
Illinois	Cook	77.0	71.02	70.67	68.95	67.86
Illinois	Du Page	65.0	60.28	59.92	58.6	57.57
Illinois	Effingham	70.0	60.04	59.64	57.4	56.7
Illinois	Hamilton	70.0	59.82	59.39	56.95	56.2
Illinois	Jersey	73.7	63.53	63.02	60.41	59.2
Illinois	Kane	69.3	60.02	59.48	57.16	55.98
Illinois	Lake	75.0	60.03	59.85	57.1	56.51
Illinois	McHenry	68.3	59.11	58.65	56.56	55.54
Illinois	McLean	72.0	61.00	60.55	58.31	57.5
Illinois	Macon	71.7	61.22	60.80	58.67	57.91
Illinois	Macoupin	70.7	59.51	58.99	56.03	55.01
Illinois	Madison	79.7	67.59	66.89	63.94	62.82
Illinois	Peoria	73.3	62.92	62.43	60.31	59.43
Illinois	Randolph	71.3	61.56	61.11	58.43	57.57
Illinois	Rock Island	65.3	56.11	55.74	53.53	52.82
Illinois	St Clair	74.7	66.01	65.48	62.79	61.56
Illinois	Sangamon	66.7	56.72	56.30	54.03	53.23
Illinois	Will	67.3	58.75	58.29	56.26	55.24
Illinois	Winnebago	69.0	56.45	55.81	52.61	51.37
Indiana	Allen	73.0	60.81	60.23	57.21	56.15
Indiana	Boone	78.0	66.93	66.36	63.61	62.47
Indiana	Carroll	70.0	59.90	59.43	56.92	56.05
Indiana	Delaware	72.3	60.00	59.43	56.53	55.51
Indiana	Elkhart	73.3	62.23	61.72	58.98	57.94
Indiana	Floyd	76.3	66.69	66.28	64.25	63.48
Indiana	Greene	76.7	66.36	65.91	63.38	62.59
Indiana	Hancock	76.0	65.26	64.69	61.93	60.82
Indiana	Hendricks	73.7	62.84	62.27	59.54	58.48
Indiana	Huntington	70.7	59.81	59.28	56.44	55.4
Indiana	Jackson	73.3	61.41	60.90	58.31	57.42
Indiana	Johnson	75.3	63.71	63.20	60.47	59.58
Indiana	Lake	77.5	67.73	67.58	65.23	64.5
Indiana	La Porte	73.0	64.26	64.00	61.8	60.99
Indiana	Madison	72.0	60.51	59.96	57.24	56.16
Indiana	Marion	78.0	68.10	67.56	65.01	63.92
Indiana	Morgan	76.3	65.41	64.89	62.42	61.49
Indiana	Perry	76.7	66.96	66.58	64.19	63.5

State	County	2007 Baseline DV	2018 Reference DV	2018 Tier 3 Control DV	2030 Reference DV	2030 Tier 3 Control DV
Indiana	Porter	76.0	66.73	66.56	64.26	63.55
Indiana	Posey	71.0	61.72	61.37	58.92	58.28
Indiana	St Joseph	74.3	64.38	63.91	61.26	60.27
Indiana	Shelby	76.0	65.31	64.75	62.04	60.96
Indiana	Vanderburgh	79.0	68.33	67.93	65.23	64.53
Indiana	Vigo	69.7	60.56	60.11	57.83	56.98
Indiana	Warrick	76.3	66.61	66.25	62.19	61.51
Iowa	Bremer	65.3	56.28	55.90	53.52	52.82
Iowa	Clinton	68.7	59.08	58.67	56.13	55.38
Iowa	Harrison	68.3	59.57	59.22	56.83	56.15
Iowa	Linn	68.3	59.16	58.80	56.7	55.99
Iowa	Montgomery	65.7	56.19	55.82	53.34	52.68
Iowa	Palo Alto	59.3	51.95	51.66	49.44	48.91
Iowa	Polk	61.0	52.12	51.73	48.97	48.23
Iowa	Scott	66.7	56.17	55.71	53.27	52.44
Iowa	Story	64.0	53.67	53.19	50.25	49.36
Iowa	Van Buren	66.3	56.29	55.90	53.13	52.42
Iowa	Warren	65.0	54.89	54.47	51.52	50.76
Kansas	Johnson	70.0	60.71	60.26	57.62	56.51
Kansas	Leavenworth	72.7	61.84	61.28	58.03	56.9
Kansas	Linn	69.7	59.27	58.87	56.72	55.94
Kansas	Sedgwick	67.0	58.09	57.63	55.67	54.76
Kansas	Shawnee	66.0	56.21	55.84	53.63	52.92
Kansas	Sumner	72.7	62.84	62.39	60.4	59.56
Kansas	Trego	68.7	63.14	62.92	61.79	61.4
Kansas	Wyandotte	71.7	62.60	62.11	59.22	58.06
Kentucky	Bell	69.0	56.02	55.46	52.73	51.81
Kentucky	Boone	72.0	61.47	61.04	58.51	57.72
Kentucky	Boyd	73.7	62.35	61.90	58.93	58.14
Kentucky	Bullitt	72.7	63.60	63.21	61.08	60.26
Kentucky	Campbell	76.0	65.12	64.53	61.78	60.57
Kentucky	Carter	70.0	58.94	58.49	55.68	54.95
Kentucky	Christian	81.0	67.52	67.10	64.58	63.86
Kentucky	Daviess	77.7	68.32	67.96	65.19	64.56
Kentucky	Edmonson	74.0	61.98	61.56	59.19	58.45
Kentucky	Fayette	71.3	59.38	58.79	55.69	54.54
Kentucky	Greenup	75.3	64.13	63.68	60.51	59.72
Kentucky	Hancock	75.3	65.70	65.34	62.32	61.69

State	County	2007 Baseline DV	2018 Reference DV	2018 Tier 3 Control DV	2030 Reference DV	2030 Tier 3 Control DV
Kentucky	Hardin	76.3	65.23	64.74	62.32	61.42
Kentucky	Henderson	77.0	67.34	66.98	63.2	62.53
Kentucky	Jefferson	80.0	69.68	69.25	66.98	66.15
Kentucky	Jessamine	73.7	61.85	61.35	58	57.12
Kentucky	Livingston	71.3	61.87	61.49	58.99	58.34
Kentucky	McCracken	74.3	64.81	64.46	61.87	61.22
Kentucky	Oldham	80.7	67.15	66.48	63.24	61.97
Kentucky	Perry	72.3	61.65	61.15	58.57	57.73
Kentucky	Pike	70.3	58.77	58.19	55.4	54.48
Kentucky	Pulaski	68.7	59.03	58.63	56.26	55.55
Kentucky	Simpson	75.3	61.92	61.35	58.18	57.16
Kentucky	Trigg	75.0	62.40	61.95	59.23	58.44
Kentucky	Warren	70.3	58.27	57.86	55.22	54.48
Louisiana	Ascension	81.7	73.74	73.47	70.38	69.82
Louisiana	Bossier	75.0	63.26	62.67	59.82	58.68
Louisiana	Caddo	75.3	64.76	64.31	61.97	61.14
Louisiana	Calcasieu	76.7	68.87	68.60	66.13	65.63
Louisiana	East Baton Rouge	83.0	74.50	74.17	71.55	70.87
Louisiana	Iberville	81.3	73.06	72.75	69.23	68.61
Louisiana	Jefferson	79.3	70.38	70.10	66.03	65.5
Louisiana	Lafayette	75.0	65.01	64.60	61.65	60.93
Louisiana	Lafourche	76.0	66.81	66.51	62.51	61.95
Louisiana	Livingston	79.0	70.68	70.34	67.36	66.69
Louisiana	Ouachita	67.0	55.69	55.18	52.52	51.59
Louisiana	Pointe Coupee	82.0	74.93	74.66	72.37	71.86
Louisiana	St Bernard	70.0	61.57	61.28	57.68	57.12
Louisiana	St Charles	74.0	65.21	64.93	60.94	60.39
Louisiana	St James	74.0	66.66	66.41	62.99	62.48
Louisiana	St John The Baptis	78.0	69.59	69.31	65.53	65.01
Louisiana	West Baton Rouge	78.0	70.32	70.01	67.51	66.93
Maine	Androscoggin	72.0	61.32	60.75	57.08	56.43
Maine	Cumberland	74.3	61.51	60.81	56.8	55.97
Maine	Hancock	80.3	69.70	69.12	65.03	64.28
Maine	Kennebec	70.7	58.68	58.01	54.53	53.79
Maine	Knox	72.3	61.54	60.88	57.14	56.35
Maine	Oxford	62.7	54.62	54.24	52.18	51.64

State	County	2007 Baseline DV	2018 Reference DV	2018 Tier 3 Control DV	2030 Reference DV	2030 Tier 3 Control DV
Maine	Penobscot	66.0	58.21	57.79	55.03	54.5
Maine	Washington	64.5	56.31	55.81	52.4	51.78
Maine	York	75.0	61.47	60.82	56.62	55.83
Maryland	Anne Arundel	85.7	69.42	68.43	64.77	63.22
Maryland	Baltimore	83.3	71.97	71.52	68.57	67.72
Maryland	Calvert	78.0	64.47	63.83	59.56	58.58
Maryland	Carroll	82.3	65.89	64.82	61.19	59.37
Maryland	Cecil	89.0	73.70	72.82	68.24	67.02
Maryland	Charles	80.7	64.67	63.80	60.61	59.33
Maryland	Frederick	80.3	63.69	62.69	59.6	57.75
Maryland	Garrett	73.3	64.57	64.22	62.57	62.01
Maryland	Harford	90.7	76.59	75.87	71.07	69.99
Maryland	Kent	81.3	67.08	66.25	62.12	60.94
Maryland	Montgomery	82.7	66.92	65.63	62.14	59.74
Maryland	Prince Georges	85.3	69.05	68.01	64.5	62.63
Maryland	Washington	76.7	63.30	62.58	59.17	58.17
Maryland	Baltimore City	67.0	60.40	60.01	57.88	57.19
Massachusetts	Barnstable	79.7	64.87	64.57	60.52	60.12
Massachusetts	Berkshire	76.3	62.47	61.68	57.98	57.23
Massachusetts	Bristol	78.0	63.27	62.84	58.88	58.35
Massachusetts	Dukes	81.3	69.84	69.36	65.04	64.56
Massachusetts	Essex	81.3	65.32	64.85	61.99	61.34
Massachusetts	Hampden	88.0	72.01	71.00	66.44	65.44
Massachusetts	Hampshire	83.7	68.35	67.37	62.99	62.03
Massachusetts	Middlesex	78.7	65.13	64.29	60.48	59.65
Massachusetts	Norfolk	82.0	66.35	65.91	63.73	63.1
Massachusetts	Suffolk	75.3	61.71	61.39	58.5	58.03
Massachusetts	Worcester	82.3	67.03	66.10	61.79	60.86
Michigan	Allegan	86.7	74.90	74.44	71.42	70.33
Michigan	Benzie	76.7	67.98	67.53	64.81	63.66
Michigan	Berrien	79.3	70.10	69.74	67.2	66.22
Michigan	Cass	75.0	62.79	62.23	59.37	58.28
Michigan	Clinton	73.3	61.30	60.69	57.92	56.78
Michigan	Genesee	76.3	63.75	63.13	59.85	58.59
Michigan	Huron	74.7	64.69	64.27	61.75	60.92
Michigan	Ingham	74.3	62.20	61.60	58.86	57.7
Michigan	Kalamazoo	74.3	62.21	61.66	58.71	57.65
Michigan	Kent	78.7	65.90	65.35	62.64	61.54

State	County	2007 Baseline DV	2018 Reference DV	2018 Tier 3 Control DV	2030 Reference DV	2030 Tier 3 Control DV
Michigan	Leelanau	74.0	62.33	61.96	57.76	56.89
Michigan	Lenawee	75.7	64.91	64.45	61.85	60.9
Michigan	Macomb	82.0	75.17	74.65	71.61	70.24
Michigan	Manistee	74.5	66.11	65.70	63.07	62.04
Michigan	Mason	76.3	67.70	67.27	64.87	63.8
Michigan	Missaukee	71.3	60.10	59.63	57.13	56.28
Michigan	Muskegon	82.3	72.30	72.01	69.51	68.63
Michigan	Oakland	77.3	69.63	69.24	66.83	65.73
Michigan	Ottawa	79.7	67.31	66.72	64.02	62.92
Michigan	St Clair	79.3	69.08	68.60	65.85	64.83
Michigan	Schoolcraft	75.7	65.27	64.82	62.16	61.18
Michigan	Washtenaw	74.0	64.29	63.81	61.22	60.22
Michigan	Wayne	81.7	74.58	74.13	71.4	70.27
Minnesota	Anoka	66.0	58.57	58.13	55.55	54.56
Mississippi	Adams	71.0	62.01	61.66	58.78	58.19
Mississippi	Bolivar	72.7	61.59	61.11	58.23	57.39
Mississippi	De Soto	80.7	66.27	65.50	62.26	60.89
Mississippi	Harrison	81.0	70.12	69.81	66.08	65.47
Mississippi	Hinds	70.3	55.24	54.41	50.63	49.15
Mississippi	Jackson	77.7	66.66	66.30	62.68	62.02
Mississippi	Lauderdale	70.3	57.29	56.65	53.69	52.66
Mississippi	Lee	71.7	57.79	57.17	54.34	53.39
Missouri	Cass	72.0	59.66	59.13	56.13	55.08
Missouri	Cedar	71.7	60.32	59.89	57.51	56.75
Missouri	Clay	81.3	69.26	68.62	65.02	63.68
Missouri	Clinton	80.0	67.28	66.62	62.98	61.64
Missouri	Greene	73.0	60.99	60.36	57.34	56.15
Missouri	Jefferson	86.0	75.66	74.95	71.78	70.22
Missouri	Lincoln	81.0	69.07	68.50	65.38	64.31
Missouri	Monroe	71.0	60.74	60.29	57.72	56.94
Missouri	Perry	77.0	65.55	65.06	62.09	61.23
Missouri	St Charles	84.0	73.15	72.56	69.72	68.4
Missouri	Ste Genevieve	79.3	67.57	67.02	63.97	62.92
Missouri	St Louis	82.3	73.89	73.27	70.3	68.88
Missouri	St Louis City	83.5	73.58	72.99	69.93	68.45
Montana	Yellowstone	59.0	52.91	52.66	50.78	50.33
Nebraska	Douglas	64.3	55.74	55.43	53.13	52.53
Nebraska	Lancaster	53.3	47.09	46.85	45.11	44.67

State	County	2007 Baseline DV	2018 Reference DV	2018 Tier 3 Control DV	2030 Reference DV	2030 Tier 3 Control DV
Nevada	Churchill	65.7	58.40	58.33	56.19	56.1
Nevada	Clark	82.0	71.20	70.45	68.11	66.37
Nevada	Washoe	72.3	60.96	60.77	57.23	56.71
Nevada	White Pine	72.0	64.00	63.85	61.44	61.17
Nevada	Carson City	66.0	54.52	54.50	49.92	49.9
New Hampshire	Belknap	70.0	60.04	59.70	57.24	56.51
New Hampshire	Cheshire	69.0	56.75	56.04	52.63	51.86
New Hampshire	Coos	76.3	64.73	64.09	60.72	59.88
New Hampshire	Grafton	66.7	55.58	54.96	51.99	51.18
New Hampshire	Hillsborough	78.0	62.04	61.27	57.56	56.68
New Hampshire	Merrimack	70.0	59.09	58.36	55.18	54.13
New Hampshire	Rockingham	78.0	64.10	63.47	59.33	58.52
New Hampshire	Sullivan	68.5	56.48	55.80	52.52	51.67
New Jersey	Atlantic	77.0	65.68	65.38	61.16	60.62
New Jersey	Camden	87.5	74.75	74.00	70.28	69.51
New Jersey	Cumberland	80.7	66.00	65.31	60.95	60.1
New Jersey	Gloucester	85.7	71.82	71.13	67.43	66.62
New Jersey	Hudson	85.0	75.61	75.31	74.28	73.73
New Jersey	Hunterdon	85.3	70.43	69.64	65.47	64.55
New Jersey	Mercer	86.3	74.67	73.94	70.1	69.07
New Jersey	Middlesex	86.3	73.34	72.53	68.41	67.4
New Jersey	Monmouth	85.0	70.49	70.02	65.52	64.87
New Jersey	Morris	83.7	68.89	68.03	64.03	63.13
New Jersey	Ocean	86.3	72.24	71.31	66.93	65.95
New Jersey	Passaic	79.3	68.71	68.15	65.41	64.65
New Mexico	Bernalillo	72.0	60.24	59.61	57.39	56.27
New Mexico	Dona Ana	75.0	66.66	66.29	66.79	66.1
New Mexico	Eddy	68.0	64.00	63.85	63.34	63.06
New Mexico	Grant	62.5	55.51	55.40	54.66	54.27
New Mexico	Lea	67.3	63.20	63.03	62.56	62.25
New Mexico	Luna	59.0	52.86	52.65	53.06	52.61
New Mexico	Sandoval	71.5	61.68	61.38	59.98	59.45

State	County	2007 Baseline DV	2018 Reference DV	2018 Tier 3 Control DV	2030 Reference DV	2030 Tier 3 Control DV
New Mexico	San Juan	70.0	66.29	66.20	65.45	65.28
New York	Albany	72.3	58.54	57.80	53.99	53.16
New York	Bronx	73.3	66.10	65.77	63.61	63.05
New York	Chautauqua	83.0	75.01	74.63	73.61	72.83
New York	Chemung	69.0	58.53	58.04	55.36	54.64
New York	Dutchess	74.3	61.08	60.19	56.33	55.39
New York	Erie	81.0	72.72	72.38	70.29	69.58
New York	Essex	76.7	64.79	64.25	61.41	60.74
New York	Hamilton	70.7	61.24	60.78	58.28	57.64
New York	Herkimer	70.0	60.57	60.10	57.71	57.08
New York	Jefferson	76.0	61.74	61.58	59.85	59.6
New York	Madison	72.0	61.41	60.89	58.16	57.42
New York	Monroe	77.3	67.26	66.79	64.15	63.46
New York	New York	76.0	68.53	68.19	65.96	65.37
New York	Niagara	77.0	70.13	69.91	68.25	67.84
New York	Oneida	66.0	55.91	55.42	53.02	52.37
New York	Onondaga	73.3	61.30	60.66	57.65	56.8
New York	Orange	79.3	65.13	64.19	60.3	59.26
New York	Oswego	73.7	64.17	63.89	61.53	61.14
New York	Putnam	80.3	69.80	69.06	65.5	64.57
New York	Queens	76.7	68.21	67.75	65.04	64.42
New York	Rensselaer	74.3	59.63	58.83	54.87	54
New York	Richmond	80.7	71.03	70.51	67.15	66.44
New York	Saratoga	77.0	60.84	59.91	55.4	54.43
New York	Schenectady	67.0	53.36	52.58	48.75	47.9
New York	Steuben	69.5	59.74	59.29	56.83	56.15
New York	Suffolk	88.0	79.00	78.52	74.98	74.35
New York	Ulster	72.3	61.36	60.79	57.76	57.06
New York	Wayne	70.0	62.58	62.34	60.28	59.94
New York	Westchester	86.3	78.92	78.55	76.69	76.05
North Carolina	Alexander	75.0	59.96	59.10	55.65	54.25
North Carolina	Avery	67.0	53.98	53.35	50.53	49.57
North Carolina	Buncombe	71.3	57.97	57.24	53.77	52.53
North Carolina	Caldwell	74.0	57.69	56.72	52.7	51.03
North Carolina	Caswell	77.3	60.21	59.25	55.33	53.71
North Carolina	Chatham	71.7	56.25	55.41	51.77	50.42
North Carolina	Cumberland	77.7	61.77	60.82	56.45	54.87
North Carolina	Davie	81.0	65.79	64.95	61.42	59.94

State	County	2007 Baseline DV	2018 Reference DV	2018 Tier 3 Control DV	2030 Reference DV	2030 Tier 3 Control DV
North Carolina	Durham	74.0	57.66	56.68	52.54	50.9
North Carolina	Edgecombe	75.3	59.23	58.34	54.48	53.02
North Carolina	Forsyth	79.0	62.77	61.81	58.03	56.48
North Carolina	Franklin	76.3	57.98	56.90	52.89	50.93
North Carolina	Graham	78.0	62.60	61.86	58.25	56.97
North Carolina	Granville	79.3	61.40	60.33	55.89	54.04
North Carolina	Guilford	81.0	62.71	61.56	57.31	55.36
North Carolina	Haywood	77.0	61.44	60.77	57.47	56.42
North Carolina	Jackson	76.0	60.07	59.33	55.89	54.67
North Carolina	Johnston	75.0	57.75	56.69	52.6	50.74
North Carolina	Lenoir	73.7	58.87	58.11	54.41	53.27
North Carolina	Lincoln	80.3	64.60	63.71	60.33	58.77
North Carolina	Martin	72.7	60.39	59.73	56.41	55.4
North Carolina	Mecklenburg	91.0	73.45	72.36	69.2	67.13
North Carolina	New Hanover	72.0	59.76	59.16	55.94	55.08
North Carolina	Person	76.0	61.90	61.18	58.03	56.84
North Carolina	Pitt	77.0	61.29	60.52	56.84	55.62
North Carolina	Rockingham	78.7	63.04	62.14	58.56	57.23
North Carolina	Rowan	87.0	69.88	68.87	65.51	63.86
North Carolina	Swain	65.0	52.61	52.04	49.3	48.39
North Carolina	Union	79.0	62.52	61.50	58.22	56.3
North Carolina	Wake	79.0	61.12	60.03	55.98	53.93
North Carolina	Yancey	77.0	61.98	61.22	58.08	57.11
Ohio	Allen	75.5	63.41	62.81	59.66	58.61
Ohio	Ashtabula	84.7	70.12	69.87	66.48	65.81
Ohio	Athens	72.0	59.37	58.89	55.87	55.09
Ohio	Butler	83.0	70.94	70.29	67.37	66.18
Ohio	Clark	76.7	64.06	63.45	60.52	59.34
Ohio	Clermont	78.3	66.17	65.50	62.71	61.55
Ohio	Clinton	79.0	65.90	65.26	62.13	60.96
Ohio	Cuyahoga	79.0	65.54	65.57	63.69	63.66
Ohio	Delaware	76.3	64.03	63.36	59.84	58.5
Ohio	Franklin	84.0	71.46	70.71	66.83	65.26
Ohio	Geauga	73.3	63.22	62.81	60.47	59.62
Ohio	Greene	76.7	64.17	63.53	60.54	59.36
Ohio	Hamilton	84.3	72.27	71.57	68.35	66.93
Ohio	Jefferson	76.3	65.64	65.24	63.09	62.43
Ohio	Knox	76.0	62.73	61.99	58.2	56.8

State	County	2007 Baseline DV	2018 Reference DV	2018 Tier 3 Control DV	2030 Reference DV	2030 Tier 3 Control DV
Ohio	Lake	78.7	63.75	63.67	61.93	61.45
Ohio	Lawrence	74.0	63.02	62.58	59.47	58.69
Ohio	Licking	74.7	62.41	61.69	58.15	56.74
Ohio	Lorain	73.0	59.77	59.53	57.17	56.75
Ohio	Lucas	79.0	67.31	66.98	64.47	63.73
Ohio	Madison	76.7	63.21	62.56	59.28	58.16
Ohio	Mahoning	75.0	63.22	62.61	59.99	58.95
Ohio	Medina	73.0	62.10	61.58	58.71	57.8
Ohio	Miami	72.7	60.14	59.48	56.42	55.28
Ohio	Montgomery	75.0	63.88	63.30	60.56	59.47
Ohio	Portage	76.0	63.86	63.23	60.14	58.91
Ohio	Preble	72.0	60.86	60.33	57.49	56.52
Ohio	Stark	78.7	65.21	64.56	61.32	60.13
Ohio	Summit	82.3	69.79	69.03	65.45	63.96
Ohio	Trumbull	80.3	67.55	66.90	63.99	62.83
Ohio	Warren	85.0	71.63	70.91	67.73	66.44
Ohio	Washington	81.0	68.33	67.92	65.53	64.85
Ohio	Wood	76.3	65.04	64.51	61.41	60.44
Oklahoma	Adair	72.3	62.30	61.89	59.79	59.06
Oklahoma	Canadian	73.7	62.58	61.89	60.37	59.02
Oklahoma	Cherokee	72.3	59.41	59.03	57.29	56.59
Oklahoma	Cleveland	72.3	63.06	62.56	60.65	59.58
Oklahoma	Creek	74.3	62.05	61.38	59.27	58.12
Oklahoma	Dewey	70.0	62.30	61.95	60.44	59.8
Oklahoma	Kay	74.0	62.51	62.05	60.12	59.25
Oklahoma	Mc Clain	70.0	60.48	60.00	58.16	57.23
Oklahoma	Mayes	73.0	61.56	61.22	59.75	59.07
Oklahoma	Oklahoma	78.0	64.45	63.68	61.41	60.09
Oklahoma	Ottawa	72.0	60.73	60.32	58.43	57.68
Oklahoma	Pittsburg	70.7	61.26	60.85	58.93	58.15
Oklahoma	Sequoyah	66.0	56.85	56.44	54.45	53.69
Oklahoma	Tulsa	77.7	66.44	65.84	63.81	62.52
Oregon	Clackamas	64.3	59.43	59.06	54.02	53.52
Oregon	Jackson	67.3	55.71	55.04	48.4	47.85
Oregon	Lane	63.7	55.58	55.06	47.97	47.34
Oregon	Marion	66.0	57.69	57.16	50.47	49.85
Oregon	Umatilla	63.0	57.60	57.31	50.14	49.82
Pennsylvania	Adams	76.0	63.18	62.43	59.24	58.07

State	County	2007 Baseline DV	2018 Reference DV	2018 Tier 3 Control DV	2030 Reference DV	2030 Tier 3 Control DV
Pennsylvania	Allegheny	85.0	73.38	72.91	70.29	69.58
Pennsylvania	Armstrong	80.0	68.00	67.48	64.34	63.65
Pennsylvania	Beaver	77.7	68.77	68.39	66.34	65.71
Pennsylvania	Berks	79.0	66.91	66.29	62.87	62.08
Pennsylvania	Blair	71.7	62.23	61.84	59.43	58.84
Pennsylvania	Bucks	90.7	78.71	77.87	73.39	72.34
Pennsylvania	Cambria	70.3	60.49	60.13	58.32	57.78
Pennsylvania	Centre	74.3	64.18	63.71	61.07	60.43
Pennsylvania	Chester	81.3	66.53	65.69	61.19	60.01
Pennsylvania	Clearfield	73.3	62.42	61.98	59.51	58.9
Pennsylvania	Dauphin	78.0	67.84	67.32	64.39	63.73
Pennsylvania	Delaware	81.7	68.49	67.78	63.89	62.98
Pennsylvania	Erie	78.3	69.42	69.09	66.48	65.74
Pennsylvania	Franklin	71.0	59.17	58.51	55.38	54.37
Pennsylvania	Greene	76.0	63.66	63.22	60.94	60.27
Pennsylvania	Indiana	76.3	66.08	65.67	63.43	62.87
Pennsylvania	Lackawanna	73.7	61.51	60.92	57.79	57.08
Pennsylvania	Lancaster	81.0	69.20	68.63	65.39	64.69
Pennsylvania	Lawrence	71.0	59.63	59.09	56.7	55.84
Pennsylvania	Lehigh	79.3	66.86	66.21	62.63	61.85
Pennsylvania	Luzerne	73.7	62.28	61.78	58.87	58.25
Pennsylvania	Lycoming	76.0	64.56	63.98	60.29	59.6
Pennsylvania	Mercer	80.0	67.26	66.61	63.88	62.75
Pennsylvania	Monroe	72.5	59.70	59.04	55.58	54.77
Pennsylvania	Montgomery	83.0	71.66	70.99	67.52	66.57
Pennsylvania	Northampton	78.3	66.45	65.83	62.48	61.69
Pennsylvania	Perry	75.3	63.70	63.16	59.99	59.25
Pennsylvania	Philadelphia	88.0	78.12	77.32	73.02	71.99
Pennsylvania	Tioga	72.7	61.69	61.16	58.09	57.42
Pennsylvania	Washington	75.3	66.33	65.95	63.92	63.33
Pennsylvania	Westmoreland	75.3	65.31	64.91	62.62	61.99
Pennsylvania	York	80.0	67.56	66.94	63.9	63.14
Rhode Island	Kent	81.0	67.79	67.14	63.19	62.49
Rhode Island	Providence	81.0	65.97	65.63	61.43	60.93
Rhode Island	Washington	80.7	67.65	67.19	62.92	62.38
South Carolina	Abbeville	77.0	61.41	60.56	56.83	55.48
South Carolina	Aiken	76.0	60.99	60.09	55.85	54.49
South Carolina	Barnwell	73.0	59.71	58.91	55.16	53.96

State	County	2007 Baseline DV	2018 Reference DV	2018 Tier 3 Control DV	2030 Reference DV	2030 Tier 3 Control DV
South Carolina	Berkeley	62.3	51.05	50.43	48.25	47.22
South Carolina	Charleston	71.0	58.48	57.84	54.03	53
South Carolina	Chester	76.0	62.67	61.90	59.21	57.84
South Carolina	Chesterfield	72.7	58.71	58.02	55.18	54.04
South Carolina	Colleton	71.3	57.64	56.98	54.02	52.95
South Carolina	Darlington	74.0	59.59	58.87	55.68	54.61
South Carolina	Edgefield	69.7	55.04	54.26	50.55	49.34
South Carolina	Pickens	78.7	60.95	59.98	55.94	54.35
South Carolina	Richland	78.7	61.97	61.00	57.25	55.53
South Carolina	Spartanburg	81.7	63.04	61.92	57.36	55.4
South Carolina	Union	77.0	61.94	61.17	58.1	56.82
South Carolina	Williamsburg	70.0	57.58	57.00	54.63	53.71
South Carolina	York	76.0	60.69	59.78	56.63	55.02
South Dakota	Custer	66.7	61.78	61.62	59.81	59.5
South Dakota	Jackson	68.0	63.46	63.32	61.76	61.49
South Dakota	Meade	56.0	51.50	51.32	49.92	49.58
South Dakota	Minnehaha	66.0	57.88	57.53	55.05	54.37
Tennessee	Anderson	76.3	58.76	57.88	53.84	52.25
Tennessee	Blount	83.3	64.84	63.86	59.7	58
Tennessee	Davidson	75.0	58.87	57.99	53.72	52.08
Tennessee	Hamilton	82.3	64.02	62.91	57.79	55.91
Tennessee	Jefferson	80.3	62.21	61.30	57.22	55.63
Tennessee	Knox	86.0	65.21	64.05	59.16	57.14
Tennessee	Loudon	77.0	60.16	59.39	55.67	54.4
Tennessee	Meigs	78.0	61.15	60.27	55.74	54.21
Tennessee	Rutherford	77.3	61.48	60.68	56.71	55.22
Tennessee	Sevier	82.0	65.76	64.96	61.24	60.01
Tennessee	Shelby	80.7	67.86	67.23	64.44	63.22
Tennessee	Sullivan	80.0	68.29	67.69	64.76	63.81
Tennessee	Sumner	82.0	65.02	64.09	59.74	58.05
Tennessee	Williamson	74.7	60.87	60.14	56.64	55.3
Tennessee	Wilson	79.3	62.43	61.59	57.23	55.67
Texas	Bexar	77.7	67.12	66.41	63.35	61.71
Texas	Brazoria	86.7	74.37	73.55	70.09	68.04
Texas	Brewster	66.0	60.76	60.54	60.08	59.71
Texas	Cameron	63.0	56.74	56.45	54.57	54.11
Texas	Collin	83.3	68.69	67.77	64.56	62.43
Texas	Dallas	82.3	70.71	69.90	67.29	65.23

State	County	2007 Baseline DV	2018 Reference DV	2018 Tier 3 Control DV	2030 Reference DV	2030 Tier 3 Control DV
Texas	Denton	90.0	76.04	75.25	72.94	71.26
Texas	Ellis	78.0	66.01	65.43	63.11	61.93
Texas	El Paso	76.3	69.02	68.66	70.17	69.46
Texas	Galveston	77.0	69.06	68.73	65.01	64.33
Texas	Gregg	79.0	71.24	70.91	69.97	69.4
Texas	Harris	90.3	78.35	77.89	75.03	73.31
Texas	Harrison	72.3	63.86	63.54	62.01	61.48
Texas	Hidalgo	63.0	55.80	55.46	53.95	53.37
Texas	Hood	80.7	69.97	69.47	67.57	66.52
Texas	Hunt	70.7	60.34	59.92	57.54	56.73
Texas	Jefferson	80.3	71.34	71.10	67.26	66.78
Texas	Johnson	83.7	72.14	71.63	69.53	68.48
Texas	Kaufman	73.0	63.78	63.32	60.97	60.08
Texas	Montgomery	78.3	67.98	67.34	64.66	63.19
Texas	Nueces	69.7	61.94	61.62	58.63	58.01
Texas	Orange	72.3	63.23	62.98	60.18	59.7
Texas	Parker	85.3	72.87	72.26	70.26	68.91
Texas	Rockwall	76.0	65.10	64.52	62.17	60.93
Texas	Smith	77.0	64.93	64.37	61.51	60.52
Texas	Tarrant	90.0	77.04	76.31	74.19	72.5
Texas	Travis	77.3	65.77	65.14	63.03	61.71
Texas	Victoria	66.7	58.00	57.62	55.34	54.64
Texas	Webb	57.7	50.82	50.53	49.25	48.76
Utah	Box Elder	75.0	67.08	66.82	64.83	64.37
Utah	Cache	70.0	62.35	62.00	59.99	59.36
Utah	Davis	80.7	73.03	72.65	69.91	69
Utah	Salt Lake	81.0	73.94	73.54	70.96	70.17
Utah	San Juan	70.3	64.81	64.73	63.71	63.57
Utah	Tooele	75.0	68.64	68.37	65.77	65.2
Utah	Utah	75.0	66.33	65.72	63.53	62.35
Utah	Washington	73.0	64.59	64.48	62	61.83
Utah	Weber	81.0	74.64	74.25	71.53	70.47
Vermont	Bennington	71.3	56.88	56.06	52.06	51.18
Vermont	Chittenden	69.7	58.29	57.66	54.42	53.68
Virginia	Arlington	83.7	70.55	69.43	66.88	64.18
Virginia	Caroline	78.7	64.15	63.31	60.66	58.94
Virginia	Charles City	80.7	69.63	69.07	66.56	65.5
Virginia	Chesterfield	76.3	65.28	64.64	62.07	60.75

State	County	2007 Baseline DV	2018 Reference DV	2018 Tier 3 Control DV	2030 Reference DV	2030 Tier 3 Control DV
Virginia	Fairfax	85.3	70.40	69.25	66.86	64.18
Virginia	Fauquier	69.7	57.88	57.25	54.51	53.4
Virginia	Frederick	71.7	56.92	56.39	53.42	52.61
Virginia	Hanover	78.7	66.41	65.75	63.03	61.76
Virginia	Henrico	82.7	70.83	70.17	67.45	66.11
Virginia	Loudoun	80.7	64.81	63.74	60.71	58.66
Virginia	Madison	74.7	63.86	63.31	60.49	59.67
Virginia	Page	69.7	59.86	59.36	56.79	56.01
Virginia	Prince William	75.7	62.14	61.33	58.51	56.85
Virginia	Roanoke	73.3	60.08	59.27	55.25	53.86
Virginia	Rockbridge	66.3	54.92	54.37	51.51	50.6
Virginia	Rockingham	67.0	57.44	56.97	54.54	53.81
Virginia	Stafford	79.3	62.67	61.63	58.83	56.35
Virginia	Wythe	69.7	58.03	57.51	54.88	54.05
Virginia	Alexandria City	79.7	65.78	64.70	62.47	59.97
Virginia	Hampton City	76.5	66.39	65.95	62.18	61.26
Virginia	Suffolk City	74.7	65.70	65.29	62.16	61.39
Washington	King	73.7	68.34	68.06	62.51	61.82
Washington	Pierce	67.3	61.55	61.18	54.99	54.37
Washington	Skagit	46.0	44.80	44.78	43.8	43.77
Washington	Spokane	62.7	53.79	53.21	47.18	46.54
West Virginia	Berkeley	73.0	58.45	57.83	54.65	53.67
West Virginia	Cabell	79.0	66.38	65.87	62.51	61.65
West Virginia	Greenbrier	70.0	59.91	59.51	56.98	56.32
West Virginia	Hancock	76.0	66.86	66.48	64.36	63.71
West Virginia	Kanawha	76.7	62.84	62.43	59.22	58.49
West Virginia	Monongalia	73.3	64.22	63.92	61.98	61.52
West Virginia	Ohio	75.5	62.92	62.52	60.49	59.81
West Virginia	Wood	77.3	64.54	64.14	61.74	61.08
Wisconsin	Ashland	61.3	53.51	53.16	50.73	50.11
Wisconsin	Brown	71.7	64.75	64.58	61.75	61.14
Wisconsin	Columbia	70.0	59.23	58.66	56.3	55.24
Wisconsin	Dane	70.3	59.52	58.89	56.42	55.15
Wisconsin	Dodge	70.0	59.45	58.92	56.35	55.28
Wisconsin	Door	82.7	71.33	70.84	68.13	66.92
Wisconsin	Florence	65.3	55.87	55.45	53.48	52.69
Wisconsin	Fond Du Lac	69.3	59.75	59.25	57.08	56.03
Wisconsin	Forest	68.0	58.52	58.11	55.76	54.97

State	County	2007 Baseline DV	2018 Reference DV	2018 Tier 3 Control DV	2030 Reference DV	2030 Tier 3 Control DV
Wisconsin	Jefferson	71.3	60.76	60.22	57.73	56.58
Wisconsin	Kenosha	79.7	64.28	64.08	61.15	60.53
Wisconsin	Kewaunee	77.0	67.71	67.35	65.07	64.03
Wisconsin	Manitowoc	78.7	68.03	67.59	64.82	63.67
Wisconsin	Marathon	67.7	58.87	58.47	56.33	55.58
Wisconsin	Milwaukee	77.3	66.19	65.96	63.29	62.58
Wisconsin	Outagamie	69.7	60.20	59.73	57.33	56.37
Wisconsin	Ozaukee	76.7	66.32	66.07	63.79	62.99
Wisconsin	Racine	74.3	61.16	61.00	58.19	57.63
Wisconsin	Rock	70.7	60.08	59.55	56.69	55.65
Wisconsin	St Croix	68.3	58.61	58.12	55.12	54.23
Wisconsin	Sauk	66.3	57.11	56.67	54.43	53.59
Wisconsin	Sheboygan	83.3	71.93	71.57	69.02	67.97
Wisconsin	Vernon	67.7	57.76	57.35	54.93	54.11
Wisconsin	Walworth	71.7	61.23	60.72	58.24	57.1
Wisconsin	Washington	67.3	59.31	58.90	56.89	55.89
Wisconsin	Waukesha	67.0	58.97	58.55	56.49	55.51
Wyoming	Campbell	68.3	63.49	63.32	61.9	61.62
Wyoming	Sublette	79.0	73.41	73.27	71.88	71.62
Wyoming	Sweetwater	64.0	59.28	59.14	58.06	57.82
Wyoming	Teton	64.7	60.31	60.19	58.92	58.73
Wyoming	Uinta	64.0	56.83	56.58	54.85	54.43

Air Quality Modeling Technical Support Document: Tier 3 Motor Vehicle Emission and Fuel Standards

Appendix C

Annual PM$_{2.5}$ Design Values for Air Quality Modeling Scenarios

U.S. Environmental Protection Agency
Office of Air Quality Planning and Standards
Air Quality Assessment Division
Research Triangle Park, NC 27711
February 2014

Table C-1. Annual PM$_{2.5}$ Design Values for Tier 3 Scenarios
(units are ug/m^3)

State	County	2007 Baseline DV	2018 Reference DV	2018 Tier 3 Control DV	2030 Reference DV	2030 Tier 3 Control DV
Alabama	Baldwin	10.80	7.52	7.51	7.42	7.39
Alabama	Clay	12.04	8.23	8.22	8.16	8.13
Alabama	Colbert	12.05	8.51	8.49	8.41	8.37
Alabama	DeKalb	12.79	8.20	8.18	8.10	8.06
Alabama	Escambia	13.18	9.80	9.79	9.70	9.67
Alabama	Etowah	13.87	9.23	9.21	9.11	9.06
Alabama	Houston	11.89	8.70	8.68	8.61	8.58
Alabama	Jefferson	17.01	11.89	11.87	11.65	11.57
Alabama	Madison	12.80	8.35	8.33	8.24	8.19
Alabama	Mobile	11.39	8.19	8.18	8.01	7.98
Alabama	Montgomery	13.70	9.95	9.94	9.84	9.80
Alabama	Morgan	12.59	8.45	8.43	8.35	8.31
Alabama	Russell	14.29	10.44	10.43	10.33	10.29
Alabama	Shelby	13.11	9.03	9.02	8.91	8.87
Alabama	Tuscaloosa	12.59	8.91	8.90	8.83	8.80
Alabama	Walker	13.06	8.99	8.98	8.88	8.85
Arizona	Cochise	6.83	6.87	6.86	7.36	7.35
Arizona	Coconino	6.93	6.67	6.67	6.76	6.73
Arizona	Gila	8.93	8.59	8.58	8.71	8.68
Arizona	Maricopa	11.98	10.86	10.86	10.88	10.80
Arizona	Pima	5.79	5.44	5.43	5.51	5.49
Arizona	Pinal	9.33	8.74	8.74	8.84	8.79
Arizona	Santa Cruz	12.67	12.55	12.54	13.24	13.21
Arkansas	Arkansas	11.82	8.92	8.91	8.80	8.78
Arkansas	Ashley	12.03	9.37	9.36	9.30	9.27
Arkansas	Crittenden	12.53	8.71	8.70	8.46	8.41
Arkansas	Faulkner	11.82	9.12	9.11	9.03	8.99
Arkansas	Garland	11.79	9.15	9.14	9.06	9.03
Arkansas	Jackson	11.19	8.42	8.41	8.32	8.30
Arkansas	Phillips	11.68	8.30	8.29	8.13	8.11
Arkansas	Polk	11.38	8.93	8.92	8.86	8.84
Arkansas	Pope	12.30	9.81	9.80	9.73	9.70
Arkansas	Pulaski	12.85	9.93	9.92	9.82	9.77
Arkansas	Sebastian	11.42	9.05	9.03	8.96	8.93
Arkansas	Union	12.02	9.41	9.40	9.33	9.31

State	County	2007 Baseline DV	2018 Reference DV	2018 Tier 3 Control DV	2030 Reference DV	2030 Tier 3 Control DV
Arkansas	White	11.54	8.77	8.76	8.68	8.66
California	Alameda	9.43	8.21	8.20	8.01	8.01
California	Butte	11.65	10.97	10.97	10.82	10.82
California	Calaveras	7.90	7.00	7.00	6.81	6.81
California	Colusa	7.89	7.33	7.33	7.22	7.21
California	Contra Costa	8.87	7.80	7.80	7.68	7.67
California	Fresno	16.88	14.20	14.19	13.39	13.39
California	Humboldt	7.38	6.98	6.98	6.96	6.96
California	Imperial	12.90	13.06	13.06	14.92	14.91
California	Inyo	6.14	5.56	5.56	5.50	5.50
California	Kern	21.20	16.98	16.98	15.85	15.84
California	Kings	17.28	14.20	14.20	13.19	13.19
California	Lake	4.84	4.56	4.56	4.53	4.53
California	Los Angeles	16.23	12.52	12.52	12.24	12.24
California	Mendocino	6.81	6.28	6.28	6.21	6.21
California	Merced	14.70	13.14	13.14	12.64	12.64
California	Monterey	6.90	6.01	6.01	5.94	5.94
California	Nevada	6.91	6.56	6.56	6.49	6.49
California	Orange	13.18	10.12	10.12	9.83	9.83
California	Placer	9.43	8.48	8.48	8.25	8.25
California	Plumas	11.48	11.12	11.12	11.02	11.02
California	Riverside	19.20	15.40	15.39	14.72	14.71
California	Sacramento	12.12	11.04	11.04	10.75	10.75
California	San Benito	6.24	5.29	5.28	5.17	5.16
California	San Bernardino	17.17	13.88	13.88	13.37	13.36
California	San Diego	12.97	10.73	10.73	10.67	10.67
California	San Francisco	9.35	7.99	7.99	7.88	7.88
California	San Joaquin	12.73	11.44	11.44	11.06	11.06
California	San Luis Obispo	8.12	6.48	6.48	6.20	6.20
California	San Mateo	8.87	7.54	7.54	7.41	7.41
California	Santa Barbara	9.98	8.98	8.98	8.93	8.93
California	Santa Clara	10.95	9.68	9.68	9.49	9.49
California	Santa Cruz	6.47	5.74	5.74	5.68	5.68
California	Shasta	6.88	6.66	6.66	6.63	6.63
California	Solano	9.81	8.82	8.82	8.66	8.66
California	Sonoma	8.24	7.57	7.57	7.47	7.47
California	Stanislaus	14.46	12.81	12.81	12.27	12.27
California	Sutter	9.16	8.34	8.34	8.12	8.12

State	County	2007 Baseline DV	2018 Reference DV	2018 Tier 3 Control DV	2030 Reference DV	2030 Tier 3 Control DV
California	Tulare	19.07	15.82	15.81	14.80	14.79
California	Ventura	10.94	8.47	8.47	8.32	8.32
California	Yolo	8.75	7.88	7.88	7.69	7.69
Colorado	Adams	9.86	8.67	8.67	8.49	8.37
Colorado	Arapahoe	7.61	6.54	6.54	6.40	6.32
Colorado	Boulder	8.13	7.42	7.42	7.30	7.24
Colorado	Denver	9.19	7.99	7.99	7.81	7.70
Colorado	Douglas	6.17	5.38	5.38	5.29	5.23
Colorado	Elbert	4.44	4.05	4.05	4.01	3.99
Colorado	El Paso	7.70	7.11	7.10	7.03	6.99
Colorado	Larimer	7.28	6.76	6.76	6.69	6.63
Colorado	Mesa	9.34	8.71	8.70	8.61	8.57
Colorado	Pueblo	7.69	7.21	7.20	7.14	7.11
Colorado	Weld	9.08	8.35	8.34	8.23	8.16
Connecticut	Fairfield	12.28	8.89	8.88	8.71	8.69
Connecticut	Hartford	10.00	7.67	7.66	7.57	7.56
Connecticut	Litchfield	8.83	6.54	6.54	6.46	6.45
Connecticut	New Haven	11.84	9.10	9.09	8.97	8.95
Connecticut	New London	10.12	7.59	7.58	7.50	7.49
Delaware	Kent	11.65	7.33	7.31	7.21	7.18
Delaware	New Castle	13.95	9.38	9.37	9.17	9.14
Delaware	Sussex	12.59	8.13	8.11	8.01	7.98
District of Co	District of Columbia	13.12	8.28	8.26	8.17	8.12
Florida	Alachua	8.66	6.30	6.29	6.24	6.21
Florida	Bay	10.55	7.75	7.74	7.66	7.63
Florida	Brevard	7.72	5.60	5.60	5.45	5.43
Florida	Broward	7.83	5.89	5.88	5.60	5.56
Florida	Citrus	8.18	5.66	5.65	5.57	5.56
Florida	Duval	9.60	7.11	7.10	7.01	6.97
Florida	Escambia	10.45	7.36	7.35	7.27	7.23
Florida	Hillsborough	9.56	6.69	6.68	6.48	6.43
Florida	Lee	7.67	5.63	5.62	5.45	5.43
Florida	Leon	11.11	8.49	8.48	8.39	8.35
Florida	Manatee	8.68	6.01	6.01	5.83	5.80
Florida	Marion	9.59	7.01	7.00	6.89	6.86
Florida	Miami-Dade	8.64	6.63	6.62	6.31	6.25
Florida	Orange	8.47	6.02	6.01	5.90	5.86

State	County	2007 Baseline DV	2018 Reference DV	2018 Tier 3 Control DV	2030 Reference DV	2030 Tier 3 Control DV
Florida	Palm Beach	7.03	5.28	5.27	5.07	5.05
Florida	Pinellas	8.90	6.08	6.07	5.92	5.88
Florida	Polk	8.63	6.15	6.15	5.98	5.95
Florida	St. Lucie	7.90	5.79	5.78	5.57	5.55
Florida	Sarasota	7.79	5.41	5.40	5.25	5.23
Florida	Seminole	8.50	6.05	6.04	5.92	5.88
Florida	Volusia	9.25	6.69	6.68	6.56	6.52
Georgia	Bibb	15.06	11.26	11.24	11.09	11.04
Georgia	Chatham	13.68	10.04	10.03	9.87	9.83
Georgia	Clarke	14.90	10.26	10.23	10.08	10.01
Georgia	Clayton	14.98	9.88	9.86	9.67	9.57
Georgia	Cobb	14.83	9.78	9.76	9.56	9.46
Georgia	DeKalb	14.25	9.11	9.08	8.89	8.79
Georgia	Dougherty	13.72	10.67	10.66	10.57	10.54
Georgia	Floyd	14.71	9.79	9.77	9.65	9.60
Georgia	Fulton	15.64	10.02	10.00	9.80	9.68
Georgia	Glynn	11.13	8.34	8.33	8.23	8.19
Georgia	Gwinnett	14.30	9.34	9.31	9.13	9.03
Georgia	Hall	12.92	8.57	8.56	8.41	8.35
Georgia	Houston	12.31	9.00	8.99	8.87	8.83
Georgia	Lowndes	11.44	8.93	8.92	8.86	8.83
Georgia	Muscogee	14.15	10.29	10.28	10.17	10.13
Georgia	Paulding	13.23	8.41	8.39	8.25	8.20
Georgia	Richmond	14.67	10.45	10.43	10.31	10.26
Georgia	Washington	13.94	10.12	10.10	10.00	9.97
Georgia	Wilkinson	15.20	11.17	11.15	11.03	10.99
Idaho	Ada	6.88	6.43	6.42	6.27	6.22
Idaho	Benewah	9.63	9.31	9.31	9.13	9.11
Idaho	Canyon	8.15	7.65	7.64	7.46	7.40
Idaho	Franklin	7.70	6.87	6.85	6.65	6.58
Idaho	Idaho	9.58	9.32	9.31	9.15	9.14
Idaho	Shoshone	11.85	11.54	11.53	11.33	11.31
Illinois	Champaign	11.94	9.14	9.12	9.01	8.95
Illinois	Cook	15.12	11.87	11.83	11.59	11.46
Illinois	DuPage	12.74	9.94	9.91	9.72	9.62
Illinois	Hamilton	12.15	8.69	8.67	8.59	8.54
Illinois	Jersey	11.97	8.91	8.89	8.74	8.69
Illinois	Kane	12.82	10.05	10.02	9.82	9.72

State	County	2007 Baseline DV	2018 Reference DV	2018 Tier 3 Control DV	2030 Reference DV	2030 Tier 3 Control DV
Illinois	Lake	10.91	8.37	8.34	8.19	8.12
Illinois	McHenry	11.33	8.79	8.76	8.59	8.52
Illinois	McLean	11.65	8.88	8.86	8.75	8.69
Illinois	Macon	12.87	9.76	9.73	9.57	9.50
Illinois	Madison	15.43	11.95	11.93	11.70	11.61
Illinois	Peoria	12.31	9.65	9.62	9.49	9.42
Illinois	Randolph	12.36	9.12	9.11	8.98	8.93
Illinois	Rock Island	11.31	9.28	9.26	9.11	9.04
Illinois	St. Clair	14.41	10.92	10.89	10.68	10.59
Illinois	Sangamon	12.21	9.32	9.29	9.14	9.08
Illinois	Will	13.03	10.16	10.13	9.92	9.82
Illinois	Winnebago	12.10	9.74	9.71	9.54	9.46
Indiana	Allen	13.46	10.08	10.05	9.91	9.83
Indiana	Clark	15.55	10.37	10.35	10.29	10.22
Indiana	Delaware	12.73	9.19	9.16	9.07	9.01
Indiana	Dubois	14.94	10.18	10.16	10.07	10.01
Indiana	Floyd	13.87	9.03	9.02	8.97	8.92
Indiana	Henry	11.74	8.42	8.40	8.33	8.27
Indiana	Howard	12.79	9.44	9.42	9.31	9.24
Indiana	Knox	13.10	9.10	9.08	8.99	8.94
Indiana	Lake	14.09	11.24	11.21	11.01	10.93
Indiana	LaPorte	12.52	9.46	9.43	9.25	9.19
Indiana	Madison	12.97	9.31	9.29	9.20	9.13
Indiana	Marion	15.00	10.78	10.75	10.63	10.53
Indiana	Porter	12.68	9.71	9.69	9.50	9.44
Indiana	St. Joseph	12.74	9.77	9.74	9.58	9.51
Indiana	Spencer	13.39	9.00	8.98	8.90	8.85
Indiana	Tippecanoe	12.61	9.21	9.19	9.05	8.99
Indiana	Vanderburgh	14.25	9.97	9.95	9.88	9.82
Indiana	Vigo	13.36	9.64	9.62	9.51	9.45
Iowa	Black Hawk	11.18	9.35	9.33	9.16	9.09
Iowa	Clinton	12.73	10.92	10.90	10.69	10.62
Iowa	Johnson	11.56	9.55	9.53	9.34	9.27
Iowa	Lee	11.41	9.36	9.34	9.16	9.10
Iowa	Linn	10.53	8.52	8.50	8.31	8.25
Iowa	Montgomery	9.72	7.79	7.77	7.60	7.56
Iowa	Muscatine	13.08	11.04	11.02	10.83	10.76
Iowa	Palo Alto	9.19	7.42	7.40	7.21	7.17

State	County	2007 Baseline DV	2018 Reference DV	2018 Tier 3 Control DV	2030 Reference DV	2030 Tier 3 Control DV
Iowa	Polk	10.18	8.06	8.04	7.83	7.76
Iowa	Pottawattamie	10.95	8.82	8.80	8.62	8.55
Iowa	Scott	13.97	11.81	11.79	11.59	11.50
Iowa	Van Buren	10.17	8.18	8.16	8.00	7.95
Iowa	Woodbury	10.40	8.77	8.75	8.59	8.53
Iowa	Wright	10.06	8.04	8.02	7.80	7.75
Kansas	Johnson	9.92	7.70	7.69	7.56	7.50
Kansas	Linn	10.14	7.97	7.95	7.84	7.81
Kansas	Sedgwick	9.66	7.93	7.92	7.84	7.79
Kansas	Shawnee	9.96	8.15	8.14	8.03	7.98
Kansas	Sumner	9.29	7.59	7.58	7.50	7.46
Kansas	Wyandotte	11.41	8.92	8.90	8.74	8.67
Kentucky	Bell	13.73	9.29	9.28	9.21	9.18
Kentucky	Boyd	13.51	8.75	8.74	8.64	8.61
Kentucky	Bullitt	14.17	9.31	9.29	9.21	9.17
Kentucky	Carter	11.58	7.73	7.72	7.66	7.63
Kentucky	Christian	13.19	8.91	8.89	8.82	8.78
Kentucky	Daviess	13.28	8.93	8.91	8.85	8.80
Kentucky	Fayette	13.48	8.40	8.38	8.29	8.23
Kentucky	Franklin	12.60	7.81	7.79	7.72	7.68
Kentucky	Hardin	13.27	8.57	8.55	8.48	8.44
Kentucky	Henderson	13.36	9.17	9.15	9.08	9.03
Kentucky	Jefferson	14.68	9.51	9.50	9.44	9.38
Kentucky	Kenton	13.27	8.60	8.58	8.48	8.42
Kentucky	McCracken	13.11	9.16	9.14	9.01	8.96
Kentucky	Madison	12.26	7.35	7.33	7.26	7.23
Kentucky	Ohio	12.78	8.68	8.67	8.61	8.57
Kentucky	Perry	13.42	8.90	8.89	8.86	8.84
Kentucky	Pike	12.61	8.25	8.24	8.22	8.20
Louisiana	Caddo Parish	11.89	9.50	9.49	9.38	9.34
Louisiana	Calcasieu Parish	9.99	7.78	7.78	7.66	7.65
Louisiana	East Baton Rouge Parish	12.27	9.36	9.35	9.12	9.09
Louisiana	Iberville Parish	12.07	9.03	9.03	8.86	8.83
Louisiana	Jefferson Parish	10.42	7.61	7.60	7.42	7.40
Louisiana	Lafayette Parish	10.17	7.58	7.57	7.49	7.47
Louisiana	Ouachita Parish	10.95	8.34	8.33	8.27	8.24
Louisiana	Rapides Parish	10.08	7.59	7.58	7.53	7.51

State	County	2007 Baseline DV	2018 Reference DV	2018 Tier 3 Control DV	2030 Reference DV	2030 Tier 3 Control DV
Louisiana	St. Bernard Parish	10.90	8.02	8.02	7.82	7.80
Louisiana	Tangipahoa Parish	11.18	8.10	8.09	8.00	7.97
Louisiana	Terrebonne Parish	9.87	7.27	7.26	7.17	7.16
Louisiana	West Baton Rouge Parish	12.71	9.80	9.79	9.55	9.52
Maine	Androscoggin	8.79	7.79	7.79	7.70	7.69
Maine	Aroostook	9.22	8.77	8.77	8.73	8.73
Maine	Cumberland	9.82	8.58	8.57	8.43	8.42
Maine	Hancock	5.11	4.20	4.20	4.16	4.16
Maine	Kennebec	8.79	7.84	7.83	7.74	7.74
Maine	Oxford	9.24	8.41	8.41	8.33	8.32
Maine	Penobscot	8.36	7.32	7.31	7.22	7.22
Maine	Piscataquis	5.55	4.73	4.73	4.70	4.69
Maryland	Anne Arundel	13.29	8.65	8.64	8.52	8.49
Maryland	Baltimore	13.54	8.80	8.78	8.65	8.63
Maryland	Cecil	11.79	7.53	7.51	7.39	7.36
Maryland	Harford	11.69	7.32	7.31	7.22	7.19
Maryland	Montgomery	11.45	7.06	7.04	6.99	6.96
Maryland	Prince George's	12.40	8.03	8.01	7.92	7.88
Maryland	Washington	12.28	7.73	7.71	7.63	7.60
Maryland	Baltimore city	14.16	9.29	9.27	9.15	9.12
Massachusetts	Berkshire	9.87	7.50	7.49	7.38	7.37
Massachusetts	Bristol	8.87	6.53	6.53	6.47	6.46
Massachusetts	Essex	9.18	7.12	7.11	7.03	7.01
Massachusetts	Hampden	11.42	9.17	9.16	9.05	9.04
Massachusetts	Middlesex	8.64	6.59	6.58	6.51	6.49
Massachusetts	Plymouth	9.39	6.87	6.87	6.77	6.76
Massachusetts	Suffolk	11.59	8.80	8.78	8.60	8.58
Massachusetts	Worcester	10.77	8.52	8.51	8.38	8.36
Michigan	Allegan	10.93	8.51	8.48	8.35	8.29
Michigan	Bay	9.90	7.48	7.46	7.36	7.31
Michigan	Berrien	10.90	8.32	8.30	8.16	8.09
Michigan	Genesee	10.68	8.25	8.23	8.13	8.07
Michigan	Ingham	11.07	8.43	8.40	8.27	8.20
Michigan	Kalamazoo	12.05	9.42	9.39	9.23	9.15
Michigan	Kent	11.78	9.26	9.23	9.07	8.98

State	County	2007 Baseline DV	2018 Reference DV	2018 Tier 3 Control DV	2030 Reference DV	2030 Tier 3 Control DV
Michigan	Macomb	11.50	8.85	8.83	8.73	8.67
Michigan	Manistee	7.41	6.07	6.06	5.99	5.95
Michigan	Missaukee	7.50	6.25	6.24	6.19	6.17
Michigan	Monroe	12.60	9.58	9.55	9.39	9.32
Michigan	Muskegon	10.57	8.15	8.13	7.99	7.93
Michigan	Oakland	12.38	9.34	9.32	9.18	9.10
Michigan	Ottawa	11.54	8.99	8.97	8.81	8.73
Michigan	St. Clair	11.08	8.89	8.88	8.80	8.76
Michigan	Washtenaw	12.40	9.69	9.67	9.53	9.45
Michigan	Wayne	15.57	11.89	11.87	11.69	11.59
Minnesota	Cass	5.74	5.10	5.09	5.04	5.02
Minnesota	Dakota	9.47	7.94	7.92	7.72	7.65
Minnesota	Hennepin	9.99	8.44	8.42	8.21	8.13
Minnesota	Mille Lacs	6.67	5.68	5.66	5.56	5.53
Minnesota	Olmsted	10.01	8.15	8.13	7.91	7.85
Minnesota	Ramsey	11.06	9.51	9.49	9.22	9.13
Minnesota	St. Louis	7.57	6.83	6.82	6.69	6.66
Minnesota	Scott	9.25	7.78	7.76	7.56	7.50
Minnesota	Stearns	8.50	7.27	7.25	7.09	7.04
Mississippi	Adams	10.79	7.89	7.88	7.78	7.75
Mississippi	Bolivar	11.80	8.65	8.64	8.54	8.52
Mississippi	DeSoto	11.92	8.22	8.21	8.05	8.01
Mississippi	Forrest	13.49	10.35	10.33	10.25	10.21
Mississippi	Grenada	10.46	7.33	7.32	7.26	7.23
Mississippi	Harrison	10.93	7.93	7.92	7.84	7.81
Mississippi	Hinds	12.27	9.02	9.00	8.91	8.86
Mississippi	Jackson	10.95	7.89	7.88	7.74	7.71
Mississippi	Jones	13.89	10.68	10.66	10.58	10.54
Mississippi	Lauderdale	12.51	9.18	9.17	9.13	9.10
Mississippi	Lee	12.31	8.70	8.69	8.61	8.57
Mississippi	Lowndes	12.38	8.84	8.83	8.78	8.75
Missouri	Buchanan	12.08	9.92	9.91	9.74	9.68
Missouri	Cass	10.38	8.06	8.04	7.90	7.86
Missouri	Clay	10.63	8.29	8.27	8.15	8.09
Missouri	Greene	11.19	8.91	8.90	8.82	8.79
Missouri	Jackson	12.00	9.48	9.46	9.30	9.22
Missouri	Jefferson	13.89	10.51	10.48	10.36	10.28
Missouri	St. Charles	13.30	10.02	10.00	9.78	9.72

State	County	2007 Baseline DV	2018 Reference DV	2018 Tier 3 Control DV	2030 Reference DV	2030 Tier 3 Control DV
Missouri	Ste. Genevieve	12.75	9.65	9.64	9.53	9.49
Missouri	St. Louis	12.85	9.56	9.54	9.41	9.34
Missouri	St. Louis city	14.08	10.53	10.51	10.29	10.21
Montana	Cascade	6.02	5.92	5.92	5.87	5.86
Montana	Flathead	9.71	9.39	9.39	9.27	9.24
Montana	Gallatin	8.63	8.40	8.41	8.32	8.29
Montana	Lewis and Clark	8.42	8.29	8.30	8.24	8.22
Montana	Lincoln	13.53	13.12	13.12	12.82	12.80
Montana	Missoula	9.82	9.37	9.36	9.19	9.15
Montana	Ravalli	9.10	8.92	8.92	8.84	8.83
Montana	Sanders	7.07	6.92	6.92	6.84	6.83
Montana	Silver Bow	11.14	10.87	10.87	10.78	10.75
Montana	Yellowstone	7.68	7.53	7.53	7.45	7.43
Nebraska	Douglas	9.59	7.51	7.49	7.33	7.28
Nebraska	Hall	7.81	6.48	6.47	6.36	6.33
Nebraska	Lancaster	8.26	6.47	6.46	6.34	6.30
Nebraska	Sarpy	9.46	7.42	7.40	7.24	7.19
Nebraska	Scotts Bluff	6.29	5.74	5.74	5.61	5.59
Nebraska	Washington	8.77	6.91	6.89	6.72	6.69
Nevada	Clark	9.43	8.75	8.75	8.60	8.54
Nevada	Washoe	8.49	7.82	7.82	7.66	7.61
New Hampshire	Belknap	6.77	5.46	5.46	5.40	5.38
New Hampshire	Cheshire	11.02	9.61	9.61	9.52	9.50
New Hampshire	Grafton	7.80	6.79	6.78	6.73	6.71
New Hampshire	Hillsborough	9.57	7.84	7.83	7.76	7.73
New Hampshire	Merrimack	9.28	7.82	7.82	7.75	7.71
New Hampshire	Rockingham	8.45	7.18	7.18	7.09	7.07
New Hampshire	Sullivan	9.31	8.25	8.25	8.18	8.17
New Jersey	Atlantic	10.82	7.53	7.53	7.48	7.46
New Jersey	Bergen	12.24	8.11	8.09	7.85	7.82
New Jersey	Camden	13.40	9.27	9.26	9.10	9.07
New Jersey	Essex	13.29	8.79	8.77	8.48	8.45
New Jersey	Gloucester	11.38	7.54	7.53	7.38	7.36
New Jersey	Hudson	13.57	9.43	9.41	9.14	9.11
New Jersey	Mercer	11.74	8.07	8.06	7.93	7.91
New Jersey	Middlesex	11.27	7.68	7.66	7.52	7.50
New Jersey	Morris	10.43	6.87	6.85	6.74	6.72
New Jersey	Ocean	10.14	6.74	6.72	6.63	6.62

State	County	2007 Baseline DV	2018 Reference DV	2018 Tier 3 Control DV	2030 Reference DV	2030 Tier 3 Control DV
New Jersey	Passaic	12.17	8.06	8.04	7.81	7.78
New Jersey	Union	13.56	9.01	8.99	8.67	8.65
New Jersey	Warren	11.81	8.17	8.16	8.02	8.00
New Mexico	Bernalillo	6.61	6.02	6.01	6.03	5.99
New Mexico	Chaves	6.47	6.15	6.14	6.43	6.42
New Mexico	Doña Ana	10.36	10.11	10.11	11.21	11.15
New Mexico	Grant	5.01	4.87	4.86	5.09	5.08
New Mexico	Sandoval	7.81	7.36	7.36	7.42	7.39
New Mexico	San Juan	5.82	5.60	5.60	5.63	5.61
New Mexico	Santa Fe	4.62	4.30	4.29	4.36	4.33
New York	Albany	9.26	6.98	6.97	6.82	6.81
New York	Bronx	14.58	10.06	10.04	9.74	9.71
New York	Chautauqua	8.88	6.10	6.09	6.01	5.99
New York	Erie	11.43	8.31	8.30	8.15	8.13
New York	Essex	5.27	4.07	4.07	4.05	4.05
New York	Kings	13.01	8.90	8.88	8.64	8.62
New York	Monroe	9.64	6.74	6.73	6.63	6.61
New York	Nassau	10.86	7.31	7.30	7.15	7.13
New York	New York	15.86	11.25	11.23	10.92	10.89
New York	Niagara	10.62	7.92	7.91	7.83	7.81
New York	Onondaga	9.03	6.65	6.64	6.57	6.56
New York	Orange	10.03	6.82	6.81	6.70	6.68
New York	Queens	11.25	7.62	7.60	7.46	7.44
New York	Richmond	12.43	8.33	8.31	8.01	7.99
New York	St. Lawrence	6.22	4.88	4.87	4.84	4.83
New York	Steuben	8.15	5.56	5.55	5.53	5.52
New York	Suffolk	10.06	6.65	6.64	6.52	6.50
New York	Westchester	11.16	7.34	7.33	7.16	7.14
North Carolina	Alamance	12.73	7.74	7.72	7.66	7.61
North Carolina	Buncombe	11.22	7.27	7.26	7.17	7.13
North Carolina	Caswell	12.01	7.16	7.15	7.09	7.06
North Carolina	Catawba	13.98	9.03	9.01	8.90	8.84
North Carolina	Chatham	11.24	6.93	6.92	6.86	6.82
North Carolina	Cumberland	12.74	8.41	8.40	8.32	8.28
North Carolina	Davidson	14.15	9.03	9.01	8.92	8.86
North Carolina	Duplin	10.31	6.55	6.54	6.52	6.50
North Carolina	Durham	13.39	8.69	8.68	8.56	8.50
North Carolina	Edgecombe	11.55	7.47	7.46	7.39	7.36

State	County	2007 Baseline DV	2018 Reference DV	2018 Tier 3 Control DV	2030 Reference DV	2030 Tier 3 Control DV
North Carolina	Forsyth	13.02	7.98	7.97	7.88	7.82
North Carolina	Gaston	13.14	8.36	8.35	8.25	8.20
North Carolina	Guilford	11.28	6.67	6.66	6.60	6.56
North Carolina	Haywood	13.00	9.30	9.29	9.22	9.17
North Carolina	Jackson	11.47	7.76	7.75	7.69	7.66
North Carolina	Lenoir	10.33	6.54	6.53	6.51	6.49
North Carolina	McDowell	12.92	8.77	8.76	8.68	8.64
North Carolina	Martin	10.14	6.62	6.61	6.60	6.58
North Carolina	Mecklenburg	13.73	8.86	8.85	8.73	8.65
North Carolina	Mitchell	11.90	7.98	7.97	7.93	7.90
North Carolina	Montgomery	11.59	7.34	7.33	7.29	7.26
North Carolina	New Hanover	9.68	6.19	6.19	6.15	6.13
North Carolina	Onslow	10.48	6.63	6.62	6.61	6.59
North Carolina	Orange	12.90	7.99	7.98	7.88	7.82
North Carolina	Pitt	11.18	7.28	7.27	7.24	7.21
North Carolina	Robeson	12.09	8.08	8.07	8.02	7.98
North Carolina	Rowan	13.28	8.58	8.57	8.50	8.45
North Carolina	Swain	11.98	8.12	8.11	8.05	8.01
North Carolina	Wake	12.46	8.16	8.15	8.06	7.99
North Carolina	Watauga	10.75	6.54	6.53	6.49	6.46
North Carolina	Wayne	11.97	7.98	7.97	7.91	7.88
North Dakota	Billings	4.66	4.34	4.33	4.28	4.27
North Dakota	Burleigh	6.77	6.16	6.15	6.03	6.01
North Dakota	Cass	7.85	6.94	6.93	6.79	6.75
North Dakota	Mercer	6.28	5.82	5.82	5.73	5.72
Ohio	Athens	11.78	7.28	7.27	7.23	7.21
Ohio	Butler	14.96	10.32	10.30	10.19	10.12
Ohio	Clark	13.83	9.86	9.83	9.70	9.63
Ohio	Clermont	13.07	8.61	8.59	8.48	8.42
Ohio	Cuyahoga	15.86	11.18	11.16	11.02	10.93
Ohio	Franklin	13.84	9.58	9.55	9.43	9.35
Ohio	Greene	12.65	8.59	8.57	8.45	8.39
Ohio	Hamilton	16.00	11.02	11.00	10.83	10.73
Ohio	Jefferson	14.80	9.42	9.41	9.34	9.30
Ohio	Lake	12.28	8.43	8.42	8.31	8.27
Ohio	Lawrence	15.44	10.02	10.01	9.90	9.87
Ohio	Lorain	12.10	8.68	8.67	8.59	8.54
Ohio	Lucas	13.88	10.46	10.44	10.24	10.16

State	County	2007 Baseline DV	2018 Reference DV	2018 Tier 3 Control DV	2030 Reference DV	2030 Tier 3 Control DV
Ohio	Mahoning	13.79	9.50	9.48	9.38	9.32
Ohio	Medina	11.94	8.27	8.25	8.16	8.12
Ohio	Montgomery	14.48	9.92	9.90	9.76	9.68
Ohio	Portage	12.82	8.69	8.67	8.57	8.52
Ohio	Preble	12.92	8.92	8.89	8.79	8.73
Ohio	Scioto	13.55	8.83	8.81	8.75	8.71
Ohio	Stark	16.11	11.14	11.12	10.99	10.92
Ohio	Summit	14.22	10.03	10.01	9.88	9.80
Ohio	Trumbull	13.90	9.61	9.59	9.49	9.44
Ohio	Warren	12.53	8.46	8.44	8.34	8.28
Oklahoma	Caddo	8.60	7.04	7.03	7.08	7.05
Oklahoma	Cherokee	12.28	9.92	9.90	9.83	9.80
Oklahoma	Kay	10.29	8.70	8.69	8.64	8.61
Oklahoma	Mayes	11.62	9.23	9.22	9.11	9.08
Oklahoma	Muskogee	11.68	9.32	9.30	9.21	9.18
Oklahoma	Oklahoma	10.21	8.06	8.04	8.02	7.97
Oklahoma	Ottawa	11.26	8.99	8.98	8.87	8.83
Oklahoma	Pittsburg	11.16	8.93	8.92	8.85	8.82
Oklahoma	Sequoyah	12.07	9.67	9.65	9.57	9.54
Oklahoma	Tulsa	11.47	9.09	9.08	9.00	8.95
Oregon	Harney	9.68	9.80	9.80	9.68	9.67
Oregon	Jackson	9.96	9.84	9.83	9.63	9.62
Oregon	Josephine	8.69	8.69	8.68	8.55	8.55
Oregon	Klamath	11.49	11.62	11.62	11.47	11.46
Oregon	Lake	9.99	9.90	9.90	9.80	9.80
Oregon	Lane	11.15	10.93	10.92	10.78	10.77
Oregon	Multnomah	8.60	8.06	8.06	7.74	7.74
Oregon	Umatilla	7.97	7.78	7.77	7.53	7.52
Oregon	Union	7.54	7.37	7.37	7.19	7.18
Oregon	Washington	8.59	8.32	8.32	8.08	8.08
Pennsylvania	Adams	12.00	7.54	7.53	7.46	7.43
Pennsylvania	Allegheny	18.36	12.05	12.04	11.75	11.73
Pennsylvania	Beaver	15.19	10.52	10.51	10.42	10.38
Pennsylvania	Berks	13.06	9.07	9.05	8.90	8.86
Pennsylvania	Bucks	12.65	8.77	8.75	8.63	8.61
Pennsylvania	Cambria	14.35	9.48	9.47	9.39	9.36
Pennsylvania	Centre	11.42	7.28	7.27	7.21	7.19
Pennsylvania	Cumberland	13.24	8.83	8.81	8.67	8.64

State	County	2007 Baseline DV	2018 Reference DV	2018 Tier 3 Control DV	2030 Reference DV	2030 Tier 3 Control DV
Pennsylvania	Dauphin	13.86	9.17	9.15	8.98	8.95
Pennsylvania	Delaware	14.24	9.79	9.78	9.59	9.57
Pennsylvania	Erie	11.57	8.27	8.26	8.17	8.14
Pennsylvania	Lackawanna	10.77	7.46	7.45	7.35	7.33
Pennsylvania	Lancaster	14.73	9.88	9.85	9.68	9.64
Pennsylvania	Mercer	12.31	8.29	8.28	8.20	8.16
Pennsylvania	Montgomery	11.99	7.98	7.96	7.83	7.81
Pennsylvania	Northampton	12.89	9.11	9.10	8.94	8.92
Pennsylvania	Philadelphia	12.97	8.98	8.97	8.81	8.79
Pennsylvania	Washington	14.52	8.94	8.93	8.78	8.76
Pennsylvania	Westmoreland	14.45	8.86	8.85	8.77	8.75
Pennsylvania	York	14.77	9.84	9.82	9.65	9.62
Rhode Island	Kent	7.54	5.39	5.38	5.35	5.34
Rhode Island	Providence	11.27	8.91	8.90	8.80	8.79
South Carolina	Beaufort	11.39	7.71	7.70	7.64	7.62
South Carolina	Charleston	10.99	7.45	7.45	7.39	7.36
South Carolina	Chesterfield	11.75	7.91	7.90	7.86	7.83
South Carolina	Edgefield	12.30	8.39	8.37	8.30	8.26
South Carolina	Florence	12.32	8.17	8.16	8.10	8.06
South Carolina	Greenville	14.74	10.14	10.12	9.93	9.85
South Carolina	Greenwood	13.52	9.09	9.07	8.96	8.92
South Carolina	Horry	11.92	8.02	8.01	7.95	7.93
South Carolina	Lexington	13.46	8.94	8.93	8.84	8.78
South Carolina	Oconee	10.32	6.60	6.59	6.52	6.49
South Carolina	Richland	13.38	8.76	8.75	8.65	8.60
South Carolina	Spartanburg	13.08	8.59	8.58	8.46	8.40
South Dakota	Brookings	8.66	7.24	7.23	7.06	7.02
South Dakota	Brown	8.07	7.10	7.09	6.94	6.92
South Dakota	Codington	9.45	8.21	8.19	8.03	7.99
South Dakota	Custer	5.55	5.27	5.27	5.25	5.25
South Dakota	Jackson	5.22	4.85	4.85	4.83	4.83
South Dakota	Minnehaha	9.64	8.00	7.98	7.77	7.71
South Dakota	Pennington	8.19	7.85	7.84	7.81	7.78
Tennessee	Blount	13.89	9.69	9.68	9.57	9.50
Tennessee	Davidson	14.04	9.54	9.52	9.37	9.29
Tennessee	Dyer	11.57	7.97	7.96	7.84	7.81
Tennessee	Hamilton	13.95	9.27	9.25	9.11	9.04
Tennessee	Knox	15.71	10.54	10.53	10.33	10.25

State	County	2007 Baseline DV	2018 Reference DV	2018 Tier 3 Control DV	2030 Reference DV	2030 Tier 3 Control DV
Tennessee	Lawrence	11.18	7.67	7.66	7.61	7.58
Tennessee	Loudon	14.76	10.37	10.36	10.22	10.16
Tennessee	McMinn	13.89	9.39	9.38	9.26	9.21
Tennessee	Madison	11.17	7.63	7.62	7.54	7.51
Tennessee	Maury	12.22	8.15	8.13	8.05	8.01
Tennessee	Montgomery	12.67	8.46	8.45	8.38	8.34
Tennessee	Putnam	11.26	7.18	7.16	7.10	7.07
Tennessee	Roane	13.86	9.28	9.26	9.13	9.08
Tennessee	Shelby	13.57	9.40	9.39	9.15	9.08
Tennessee	Sullivan	13.24	8.40	8.39	8.33	8.30
Tennessee	Sumner	12.65	8.33	8.31	8.19	8.14
Texas	Bowie	12.19	9.57	9.56	9.43	9.40
Texas	Dallas	10.99	8.34	8.32	8.29	8.20
Texas	Ector	8.13	7.69	7.69	8.12	8.10
Texas	El Paso	11.21	11.22	11.21	12.76	12.70
Texas	Harris	15.04	11.99	11.97	11.93	11.84
Texas	Harrison	11.01	8.46	8.45	8.36	8.33
Texas	Hidalgo	10.94	9.57	9.57	9.67	9.64
Texas	Nueces	10.71	8.66	8.66	8.58	8.55
Texas	Orange	11.29	8.93	8.93	8.86	8.85
Texas	Potter	6.17	5.45	5.44	5.55	5.53
Texas	Tarrant	11.32	8.76	8.74	8.73	8.64
Texas	Travis	9.06	7.05	7.04	7.05	7.02
Utah	Box Elder	8.28	7.34	7.31	7.08	7.01
Utah	Cache	9.79	8.84	8.81	8.57	8.48
Utah	Davis	10.25	9.28	9.26	9.05	8.95
Utah	Salt Lake	11.69	10.66	10.64	10.41	10.29
Utah	Tooele	6.84	6.24	6.22	6.08	6.04
Utah	Utah	10.42	9.26	9.23	8.97	8.86
Utah	Weber	10.54	9.49	9.47	9.19	9.09
Vermont	Bennington	7.67	5.84	5.84	5.78	5.77
Vermont	Chittenden	8.39	7.15	7.14	7.06	7.05
Vermont	Rutland	10.67	9.66	9.65	9.57	9.56
Virginia	Arlington	12.93	8.14	8.13	8.06	8.01
Virginia	Charles City	11.35	7.23	7.21	7.16	7.12
Virginia	Chesterfield	12.30	7.78	7.77	7.72	7.66
Virginia	Fairfax	13.47	8.35	8.34	8.28	8.22
Virginia	Henrico	12.03	7.57	7.56	7.51	7.46

State	County	2007 Baseline DV	2018 Reference DV	2018 Tier 3 Control DV	2030 Reference DV	2030 Tier 3 Control DV
Virginia	Loudoun	12.17	7.57	7.56	7.50	7.46
Virginia	Page	11.71	7.08	7.06	7.04	7.01
Virginia	Rockingham	11.66	7.76	7.75	7.71	7.68
Virginia	Bristol city	12.60	7.84	7.83	7.78	7.74
Virginia	Hampton city	11.64	7.40	7.39	7.30	7.26
Virginia	Lynchburg city	11.78	7.69	7.68	7.62	7.58
Virginia	Norfolk city	12.13	8.04	8.03	7.94	7.89
Virginia	Roanoke city	13.96	9.32	9.31	9.22	9.17
Virginia	Virginia Beach city	11.56	7.55	7.54	7.47	7.43
Washington	King	9.27	8.19	8.19	7.88	7.88
Washington	Pierce	9.89	9.23	9.23	8.95	8.95
Washington	Snohomish	9.06	8.47	8.47	8.26	8.26
Washington	Spokane	9.56	9.11	9.10	8.70	8.69
Washington	Yakima	9.70	9.03	9.00	8.44	8.41
West Virginia	Berkeley	14.90	10.27	10.25	10.13	10.09
West Virginia	Brooke	15.40	9.65	9.64	9.55	9.51
West Virginia	Cabell	15.35	10.34	10.33	10.20	10.16
West Virginia	Hancock	14.31	8.92	8.91	8.84	8.81
West Virginia	Harrison	13.37	8.43	8.42	8.38	8.34
West Virginia	Kanawha	15.46	9.78	9.77	9.65	9.61
West Virginia	Marion	14.44	9.34	9.33	9.27	9.23
West Virginia	Marshall	14.27	8.76	8.75	8.71	8.67
West Virginia	Monongalia	13.58	8.06	8.05	7.99	7.96
West Virginia	Ohio	13.81	8.23	8.22	8.16	8.13
West Virginia	Raleigh	12.00	7.40	7.39	7.35	7.33
West Virginia	Wood	14.58	9.63	9.62	9.54	9.50
Wisconsin	Ashland	6.16	5.42	5.42	5.36	5.34
Wisconsin	Brown	11.73	10.01	9.99	10.14	10.05
Wisconsin	Dane	12.57	10.66	10.64	10.49	10.39
Wisconsin	Dodge	11.00	9.23	9.21	9.08	9.00
Wisconsin	Forest	7.09	6.00	5.99	5.93	5.90
Wisconsin	Grant	12.27	10.45	10.43	10.23	10.16
Wisconsin	Kenosha	12.62	10.14	10.11	9.94	9.84
Wisconsin	La Crosse	11.76	10.32	10.30	10.11	10.03
Wisconsin	Manitowoc	10.67	9.02	8.99	8.96	8.89
Wisconsin	Milwaukee	14.69	12.22	12.19	12.04	11.89
Wisconsin	Outagamie	11.25	9.49	9.46	9.40	9.33

State	County	2007 Baseline DV	2018 Reference DV	2018 Tier 3 Control DV	2030 Reference DV	2030 Tier 3 Control DV
Wisconsin	Ozaukee	11.84	9.71	9.68	9.55	9.46
Wisconsin	St. Croix	10.28	8.88	8.86	8.65	8.58
Wisconsin	Sauk	10.50	8.65	8.62	8.48	8.41
Wisconsin	Taylor	8.73	7.65	7.63	7.52	7.47
Wisconsin	Vilas	6.78	5.88	5.87	5.80	5.77
Wisconsin	Waukesha	13.82	11.49	11.45	11.28	11.15
Wyoming	Campbell	5.52	5.27	5.27	5.22	5.22
Wyoming	Converse	3.73	3.48	3.48	3.45	3.44
Wyoming	Fremont	7.72	7.43	7.42	7.37	7.35
Wyoming	Laramie	4.28	3.82	3.82	3.76	3.75
Wyoming	Sheridan	9.07	8.84	8.84	8.75	8.72
Wyoming	Sublette	6.49	6.28	6.28	6.27	6.26

Air Quality Modeling Technical Support Document: Tier 3 Motor Vehicle Emission and Fuel Standards

Appendix D

24-Hour PM$_{2.5}$ Design Values for Air Quality Modeling Scenarios

U.S. Environmental Protection Agency
Office of Air Quality Planning and Standards
Air Quality Assessment Division
Research Triangle Park, NC 27711
February 2014

Table D-1. 24-hour PM$_{2.5}$ Design Values for Tier 3 Scenarios
(units are ug/m^3)

State	County	2007 Baseline DV	2018 Reference DV	2018 Tier 3 Control DV	2030 Reference DV	2030 Tier 3 Control DV
Alabama	Baldwin	23.7	15.2	15.1	15.0	15.0
Alabama	Clay	27.2	15.9	15.8	15.7	15.7
Alabama	Colbert	28.1	15.9	15.9	15.8	15.7
Alabama	DeKalb	28.4	16.0	15.9	15.8	15.7
Alabama	Escambia	28.2	20.6	20.6	20.3	20.2
Alabama	Houston	25.5	18.6	18.6	18.4	18.3
Alabama	Jefferson	36.7	26.7	26.6	26.4	26.2
Alabama	Madison	29.5	17.0	16.9	16.8	16.6
Alabama	Mobile	24.1	15.7	15.7	15.2	15.2
Alabama	Montgomery	29.1	21.5	21.5	21.2	21.0
Alabama	Morgan	28.9	15.9	15.9	15.7	15.6
Alabama	Russell	30.3	22.8	22.8	22.6	22.6
Alabama	Shelby	28.4	19.7	19.7	19.6	19.5
Alabama	Tuscaloosa	26.9	16.9	16.9	16.7	16.6
Alabama	Walker	29.6	18.1	18.0	17.8	17.8
Arizona	Cochise	12.9	13.1	13.1	13.8	13.8
Arizona	Coconino	18.7	18.2	18.2	18.4	18.3
Arizona	Gila	22.7	21.7	21.7	21.8	21.7
Arizona	Maricopa	29.0	25.3	25.2	24.9	24.6
Arizona	Pima	12.1	11.5	11.5	11.4	11.4
Arizona	Pinal	43.6	40.8	40.8	40.4	40.2
Arizona	Santa Cruz	29.2	29.3	29.3	30.7	30.5
Arkansas	Arkansas	27.3	17.8	17.8	17.7	17.6
Arkansas	Ashley	25.9	18.2	18.2	18.0	18.0
Arkansas	Crittenden	31.0	17.4	17.4	17.0	16.9
Arkansas	Faulkner	26.0	18.8	18.8	18.7	18.6
Arkansas	Garland	26.1	18.0	17.9	17.8	17.8
Arkansas	Jackson	26.1	17.5	17.5	17.3	17.2
Arkansas	Phillips	26.9	16.8	16.7	16.5	16.4
Arkansas	Polk	25.5	18.0	18.0	18.0	18.0
Arkansas	Pope	26.6	20.1	20.1	20.0	20.0
Arkansas	Pulaski	30.3	21.0	21.0	20.8	20.7
Arkansas	Sebastian	24.5	17.9	17.9	17.8	17.7
Arkansas	Union	25.7	18.8	18.8	18.7	18.6
Arkansas	White	27.9	19.4	19.3	19.1	19.0

State	County	2007 Baseline DV	2018 Reference DV	2018 Tier 3 Control DV	2030 Reference DV	2030 Tier 3 Control DV
California	Alameda	42.0	36.8	36.8	34.9	34.9
California	Butte	48.0	46.3	46.3	45.6	45.6
California	Calaveras	27.0	23.7	23.7	22.8	22.8
California	Contra Costa	36.1	31.3	31.3	29.7	29.7
California	Fresno	57.7	47.1	47.0	42.8	42.7
California	Humboldt	24.7	23.9	23.9	23.8	23.8
California	Imperial	39.0	38.2	38.2	43.3	43.2
California	Inyo	30.8	28.2	28.2	27.9	27.9
California	Kern	69.6	49.1	49.1	42.4	42.4
California	Kings	59.2	48.3	48.3	42.5	42.5
California	Lake	22.9	22.3	22.3	22.3	22.3
California	Los Angeles	43.3	35.2	35.1	33.1	33.1
California	Mendocino	19.0	18.1	18.1	17.8	17.8
California	Merced	51.6	45.6	45.6	42.7	42.7
California	Monterey	14.2	11.7	11.7	11.5	11.5
California	Nevada	27.1	25.9	25.9	25.6	25.6
California	Orange	38.8	29.2	29.2	27.5	27.5
California	Placer	28.3	25.5	25.5	24.3	24.3
California	Plumas	32.5	31.1	31.1	30.7	30.7
California	Riverside	50.7	38.2	38.2	34.9	34.9
California	Sacramento	55.1	51.3	51.3	48.9	48.9
California	San Benito	17.0	14.0	14.0	13.7	13.6
California	San Bernardino	51.7	38.9	38.9	35.3	35.3
California	San Diego	32.7	29.7	29.7	28.8	28.8
California	San Francisco	32.7	27.5	27.5	26.1	26.1
California	San Joaquin	45.4	41.3	41.3	39.3	39.2
California	San Luis Obispo	22.7	16.6	16.6	14.9	14.9
California	San Mateo	31.0	25.6	25.6	24.3	24.2
California	Santa Barbara	22.4	20.1	20.1	19.7	19.7
California	Santa Clara	40.3	34.4	34.3	32.8	32.8
California	Santa Cruz	13.4	12.0	12.0	11.8	11.8
California	Shasta	22.1	21.5	21.5	21.4	21.4
California	Solano	40.0	36.7	36.7	35.4	35.4
California	Sonoma	30.4	27.7	27.7	26.8	26.8
California	Stanislaus	53.8	46.7	46.7	43.1	43.1
California	Sutter	33.9	30.6	30.6	29.3	29.3
California	Tulare	56.5	43.2	43.2	38.6	38.6
California	Ventura	27.6	20.6	20.6	19.8	19.8

State	County	2007 Baseline DV	2018 Reference DV	2018 Tier 3 Control DV	2030 Reference DV	2030 Tier 3 Control DV
California	Yolo	33.1	31.1	31.1	30.0	30.0
Colorado	Adams	29.4	25.8	25.8	25.1	24.6
Colorado	Arapahoe	19.4	16.9	16.9	16.5	16.3
Colorado	Boulder	22.8	20.7	20.6	20.3	20.1
Colorado	Denver	25.1	21.9	21.9	21.3	20.9
Colorado	Douglas	16.6	14.4	14.4	14.1	13.9
Colorado	Elbert	13.5	12.4	12.4	12.1	12.0
Colorado	El Paso	15.8	15.6	15.6	15.5	15.4
Colorado	Larimer	18.8	18.0	18.0	17.8	17.6
Colorado	Mesa	26.1	25.1	25.1	24.7	24.5
Colorado	Pueblo	15.6	15.7	15.7	15.6	15.6
Colorado	Weld	24.1	22.5	22.4	22.0	21.7
Connecticut	Fairfield	32.9	22.9	22.9	22.5	22.5
Connecticut	Hartford	28.6	21.8	21.8	21.6	21.6
Connecticut	Litchfield	24.0	14.9	14.9	14.8	14.8
Connecticut	New Haven	32.2	23.9	23.9	23.8	23.8
Connecticut	New London	27.9	19.6	19.6	19.4	19.4
Delaware	Kent	29.4	17.8	17.7	17.5	17.4
Delaware	New Castle	34.8	23.3	23.2	22.9	22.8
Delaware	Sussex	30.3	18.3	18.2	18.0	18.0
District of Co	District of Columbia	31.2	20.4	20.4	20.3	20.3
Florida	Alachua	20.8	15.3	15.3	15.2	15.1
Florida	Bay	24.2	16.8	16.8	16.5	16.4
Florida	Brevard	20.5	15.0	15.0	14.9	14.8
Florida	Broward	19.0	14.3	14.3	14.0	14.0
Florida	Citrus	18.6	11.8	11.8	11.6	11.6
Florida	Duval	22.1	15.9	15.9	15.7	15.6
Florida	Escambia	24.0	16.2	16.2	16.0	15.8
Florida	Hillsborough	20.0	14.4	14.4	14.1	14.0
Florida	Lee	16.4	11.7	11.7	11.7	11.7
Florida	Leon	23.5	18.4	18.4	18.2	18.2
Florida	Manatee	19.2	12.8	12.8	12.4	12.4
Florida	Marion	22.5	15.9	15.8	15.5	15.3
Florida	Miami-Dade	19.2	13.1	13.1	12.8	12.7
Florida	Orange	19.6	13.2	13.2	13.0	12.9
Florida	Palm Beach	17.8	12.0	12.0	11.9	11.9
Florida	Pinellas	20.0	13.9	13.9	13.6	13.5

State	County	2007 Baseline DV	2018 Reference DV	2018 Tier 3 Control DV	2030 Reference DV	2030 Tier 3 Control DV
Florida	Polk	17.0	12.4	12.4	12.2	12.1
Florida	St. Lucie	17.7	12.4	12.4	12.2	12.2
Florida	Sarasota	17.4	12.7	12.7	12.6	12.5
Florida	Seminole	19.0	13.1	13.1	12.9	12.8
Florida	Volusia	23.9	16.7	16.7	16.5	16.4
Georgia	Bibb	33.6	25.6	25.6	25.3	25.2
Georgia	Chatham	26.7	21.1	21.0	20.8	20.7
Georgia	Clayton	30.3	19.5	19.5	19.3	19.1
Georgia	Cobb	32.2	19.5	19.5	19.2	18.9
Georgia	DeKalb	30.9	18.5	18.4	18.2	18.0
Georgia	Dougherty	33.6	29.0	29.0	28.8	28.8
Georgia	Floyd	34.9	22.4	22.3	22.1	21.9
Georgia	Fulton	33.6	19.6	19.6	19.2	18.9
Georgia	Glynn	25.0	18.5	18.5	18.2	18.1
Georgia	Gwinnett	28.4	18.0	18.0	17.6	17.4
Georgia	Hall	28.4	17.0	17.0	16.6	16.4
Georgia	Houston	30.1	23.6	23.5	23.1	22.9
Georgia	Lowndes	25.9	20.7	20.7	20.6	20.5
Georgia	Muscogee	29.5	25.0	24.9	24.9	24.8
Georgia	Paulding	32.3	18.7	18.6	18.2	18.0
Georgia	Richmond	30.8	23.0	23.0	22.8	22.6
Georgia	Washington	29.4	19.6	19.5	19.3	19.2
Georgia	Wilkinson	32.3	24.9	24.9	24.7	24.6
Idaho	Ada	22.3	20.4	20.3	19.6	19.2
Idaho	Benewah	28.6	27.7	27.7	27.2	27.1
Idaho	Canyon	28.2	25.4	25.3	24.2	23.8
Idaho	Franklin	36.7	32.2	32.0	30.2	29.5
Idaho	Idaho	28.4	27.5	27.5	27.0	27.0
Idaho	Shoshone	35.0	33.9	33.8	33.2	33.1
Illinois	Champaign	29.2	21.8	21.8	21.6	21.4
Illinois	Cook	38.9	31.3	31.1	30.8	30.4
Illinois	DuPage	32.8	25.9	25.7	25.2	24.8
Illinois	Hamilton	28.6	21.5	21.4	21.2	21.1
Illinois	Jersey	28.0	20.1	20.0	19.7	19.6
Illinois	Kane	31.1	25.9	25.7	25.1	24.7
Illinois	Lake	29.3	22.1	22.0	21.7	21.5
Illinois	LaSalle	27.5	20.9	20.8	20.3	20.1
Illinois	McHenry	28.7	21.7	21.6	21.1	20.9

State	County	2007 Baseline DV	2018 Reference DV	2018 Tier 3 Control DV	2030 Reference DV	2030 Tier 3 Control DV
Illinois	McLean	29.0	21.9	21.8	21.5	21.3
Illinois	Macon	30.6	22.7	22.6	22.3	22.1
Illinois	Madison	34.8	25.5	25.4	24.9	24.7
Illinois	Peoria	30.2	23.4	23.3	22.8	22.5
Illinois	Randolph	26.8	21.4	21.4	21.1	20.9
Illinois	Rock Island	26.7	21.0	21.0	20.7	20.5
Illinois	St. Clair	30.0	22.8	22.7	22.4	22.2
Illinois	Sangamon	29.7	21.2	21.2	20.8	20.6
Illinois	Will	33.5	25.6	25.4	24.9	24.6
Illinois	Winnebago	30.6	25.4	25.3	24.9	24.7
Indiana	Allen	32.7	24.8	24.7	24.2	23.9
Indiana	Clark	35.6	21.4	21.4	21.2	21.0
Indiana	Delaware	28.6	19.4	19.3	18.9	18.7
Indiana	Dubois	34.9	21.0	20.9	20.7	20.5
Indiana	Floyd	31.1	17.6	17.6	17.4	17.3
Indiana	Henry	26.0	17.5	17.4	17.1	16.9
Indiana	Howard	32.9	21.5	21.4	20.9	20.6
Indiana	Knox	30.7	20.7	20.6	20.4	20.3
Indiana	Lake	32.8	27.2	27.0	26.6	26.2
Indiana	LaPorte	30.7	23.0	22.9	22.6	22.4
Indiana	Madison	30.0	19.3	19.2	18.7	18.5
Indiana	Marion	37.0	24.7	24.6	24.1	23.8
Indiana	Porter	30.3	24.4	24.3	24.0	23.8
Indiana	St. Joseph	30.0	23.8	23.7	23.1	22.9
Indiana	Spencer	28.8	18.3	18.2	18.0	17.9
Indiana	Tippecanoe	30.5	20.8	20.7	20.3	20.0
Indiana	Vanderburgh	30.3	20.8	20.8	20.5	20.4
Indiana	Vigo	34.5	22.6	22.5	22.2	21.9
Iowa	Black Hawk	29.1	25.6	25.5	24.8	24.5
Iowa	Clinton	33.0	29.1	29.0	28.5	28.2
Iowa	Johnson	30.6	24.7	24.6	24.0	23.8
Iowa	Lee	26.0	22.0	21.9	21.6	21.4
Iowa	Linn	27.2	22.5	22.4	21.9	21.6
Iowa	Montgomery	23.7	18.1	18.0	17.3	17.1
Iowa	Muscatine	36.2	32.7	32.7	32.2	31.9
Iowa	Palo Alto	24.3	18.6	18.5	17.8	17.6
Iowa	Polk	26.2	21.3	21.2	20.6	20.3
Iowa	Pottawattamie	26.3	22.6	22.5	21.9	21.5

State	County	2007 Baseline DV	2018 Reference DV	2018 Tier 3 Control DV	2030 Reference DV	2030 Tier 3 Control DV
Iowa	Scott	34.6	28.7	28.7	28.3	28.0
Iowa	Van Buren	26.2	20.5	20.5	20.3	20.2
Iowa	Woodbury	28.3	22.6	22.6	22.0	21.8
Kansas	Johnson	22.5	16.5	16.4	16.1	15.9
Kansas	Linn	22.5	15.9	15.9	15.6	15.6
Kansas	Sedgwick	23.1	17.3	17.2	16.9	16.8
Kansas	Shawnee	22.8	17.2	17.2	16.9	16.8
Kansas	Sumner	21.6	16.1	16.0	15.7	15.5
Kansas	Wyandotte	24.2	18.5	18.5	18.1	17.9
Kentucky	Bell	26.9	19.7	19.6	19.4	19.3
Kentucky	Boyd	31.2	15.5	15.5	15.3	15.2
Kentucky	Bullitt	31.6	17.4	17.4	17.2	17.1
Kentucky	Carter	27.0	14.8	14.7	14.6	14.6
Kentucky	Christian	32.3	17.8	17.7	17.5	17.4
Kentucky	Daviess	30.6	18.4	18.4	18.3	18.2
Kentucky	Fayette	29.5	18.8	18.8	18.4	18.1
Kentucky	Franklin	29.5	16.6	16.6	16.3	16.2
Kentucky	Hardin	31.8	18.1	18.0	17.9	17.8
Kentucky	Henderson	29.2	18.7	18.7	18.5	18.4
Kentucky	Jefferson	35.1	19.7	19.7	19.6	19.4
Kentucky	Kenton	30.6	17.7	17.6	17.4	17.3
Kentucky	McCracken	31.9	18.7	18.7	18.5	18.3
Kentucky	Madison	27.8	16.4	16.4	16.1	16.0
Kentucky	Ohio	29.6	16.0	16.0	15.9	15.8
Kentucky	Perry	29.8	16.8	16.7	16.9	16.8
Kentucky	Pike	28.4	17.6	17.6	17.7	17.7
Kentucky	Warren	29.0	15.3	15.2	15.1	15.0
Louisiana	Caddo Parish	24.7	19.1	19.1	18.8	18.7
Louisiana	Calcasieu Parish	22.9	18.0	18.0	17.8	17.8
Louisiana	East Baton Rouge Parish	26.1	19.0	19.0	18.4	18.3
Louisiana	Iberville Parish	25.8	19.6	19.6	19.6	19.5
Louisiana	Jefferson Parish	23.2	16.7	16.7	16.4	16.3
Louisiana	Lafayette Parish	22.3	16.1	16.0	16.0	15.9
Louisiana	Ouachita Parish	25.8	18.2	18.1	17.9	17.8
Louisiana	Rapides Parish	22.5	16.0	16.0	15.9	15.9
Louisiana	St. Bernard Parish	22.0	15.8	15.8	15.4	15.4

State	County	2007 Baseline DV	2018 Reference DV	2018 Tier 3 Control DV	2030 Reference DV	2030 Tier 3 Control DV
Louisiana	Tangipahoa Parish	25.7	18.0	18.0	17.9	17.9
Louisiana	Terrebonne Parish	22.8	16.3	16.3	16.2	16.2
Louisiana	West Baton Rouge Parish	26.0	19.6	19.6	19.0	18.9
Maine	Androscoggin	23.8	21.0	21.0	20.8	20.8
Maine	Aroostook	22.3	21.1	21.1	21.1	21.1
Maine	Cumberland	21.7	18.1	18.1	17.8	17.8
Maine	Hancock	20.5	12.3	12.2	12.1	12.1
Maine	Kennebec	21.4	18.9	18.9	18.7	18.7
Maine	Oxford	22.5	20.2	20.2	20.0	20.0
Maine	Penobscot	21.4	16.6	16.6	16.3	16.3
Maine	Piscataquis	17.2	12.4	12.4	12.3	12.3
Maryland	Anne Arundel	33.1	20.3	20.2	20.1	20.0
Maryland	Baltimore	33.0	23.2	23.1	23.1	23.0
Maryland	Cecil	27.8	18.5	18.5	18.2	18.1
Maryland	Harford	28.6	17.2	17.1	17.1	17.0
Maryland	Montgomery	28.0	16.4	16.3	16.3	16.2
Maryland	Prince George's	27.5	17.2	17.1	17.1	17.0
Maryland	Washington	29.1	17.5	17.5	17.2	17.1
Maryland	Baltimore city	34.0	25.1	25.1	24.6	24.5
Massachusetts	Berkshire	27.9	20.9	20.9	20.6	20.6
Massachusetts	Bristol	24.1	16.8	16.7	16.6	16.6
Massachusetts	Essex	26.2	19.1	19.0	18.9	18.8
Massachusetts	Hampden	30.8	24.3	24.3	24.2	24.2
Massachusetts	Middlesex	21.7	13.9	13.9	13.8	13.8
Massachusetts	Plymouth	27.0	18.1	18.1	17.9	17.9
Massachusetts	Suffolk	29.2	21.4	21.3	20.9	20.9
Massachusetts	Worcester	28.2	20.1	20.0	19.8	19.7
Michigan	Allegan	30.4	22.8	22.8	22.6	22.4
Michigan	Bay	26.9	20.7	20.6	20.4	20.2
Michigan	Berrien	28.8	21.3	21.3	21.1	20.9
Michigan	Genesee	26.9	21.6	21.5	21.4	21.2
Michigan	Ingham	28.5	22.3	22.3	22.0	21.8
Michigan	Kalamazoo	28.9	23.7	23.6	23.3	23.2
Michigan	Kent	31.1	25.1	25.0	24.7	24.4
Michigan	Macomb	31.2	23.6	23.6	23.6	23.4
Michigan	Manistee	22.5	17.1	17.0	16.8	16.6

State	County	2007 Baseline DV	2018 Reference DV	2018 Tier 3 Control DV	2030 Reference DV	2030 Tier 3 Control DV
Michigan	Missaukee	22.5	16.2	16.2	16.1	16.0
Michigan	Monroe	32.4	23.5	23.4	22.9	22.7
Michigan	Muskegon	29.4	21.4	21.3	20.9	20.7
Michigan	Oakland	35.0	25.0	24.9	24.6	24.4
Michigan	Ottawa	29.7	24.3	24.2	23.9	23.7
Michigan	St. Clair	35.5	26.0	26.0	25.9	25.7
Michigan	Washtenaw	33.6	25.1	25.0	24.4	24.1
Michigan	Wayne	38.3	30.9	30.9	30.6	30.3
Minnesota	Cass	17.9	14.9	14.8	14.5	14.4
Minnesota	Dakota	25.7	22.9	22.8	22.2	21.9
Minnesota	Hennepin	27.2	24.5	24.4	23.7	23.3
Minnesota	Mille Lacs	22.2	16.6	16.5	16.2	16.1
Minnesota	Olmsted	29.7	25.9	25.7	25.0	24.7
Minnesota	Ramsey	29.8	26.7	26.6	25.7	25.4
Minnesota	St. Louis	23.6	21.1	21.0	20.6	20.4
Minnesota	Scott	24.5	20.7	20.6	19.9	19.6
Minnesota	Stearns	22.1	19.9	19.8	19.2	19.0
Minnesota	Washington	30.2	28.8	28.8	28.0	27.7
Mississippi	Adams	24.0	16.5	16.5	16.3	16.2
Mississippi	Bolivar	26.4	18.3	18.2	18.0	18.0
Mississippi	DeSoto	26.9	15.2	15.2	14.9	14.8
Mississippi	Forrest	28.4	21.5	21.5	21.2	21.2
Mississippi	Grenada	22.8	14.2	14.2	14.0	14.0
Mississippi	Harrison	24.5	16.9	16.9	16.7	16.6
Mississippi	Hinds	26.2	17.4	17.3	17.1	16.9
Mississippi	Jackson	24.7	16.6	16.5	16.3	16.2
Mississippi	Jones	28.5	21.9	21.8	21.6	21.5
Mississippi	Lauderdale	26.4	18.3	18.2	18.0	17.9
Mississippi	Lee	29.8	16.8	16.7	16.6	16.5
Mississippi	Lowndes	28.1	18.9	18.9	18.8	18.7
Missouri	Buchanan	27.0	22.2	22.1	21.6	21.4
Missouri	Cass	24.6	17.5	17.5	17.1	17.0
Missouri	Clay	24.7	18.0	18.0	17.6	17.5
Missouri	Greene	25.7	18.9	18.8	18.6	18.4
Missouri	Jackson	26.6	21.8	21.7	21.2	21.0
Missouri	Jefferson	34.2	23.8	23.8	23.7	23.5
Missouri	St. Charles	32.8	22.9	22.8	22.4	22.2
Missouri	Ste. Genevieve	29.8	20.4	20.4	20.2	20.1

State	County	2007 Baseline DV	2018 Reference DV	2018 Tier 3 Control DV	2030 Reference DV	2030 Tier 3 Control DV
Missouri	St. Louis	30.9	23.5	23.5	23.3	23.0
Missouri	St. Louis city	32.4	23.6	23.5	23.1	22.8
Montana	Cascade	17.3	17.1	17.1	17.0	16.9
Montana	Flathead	22.7	22.2	22.2	22.0	21.9
Montana	Gallatin	27.0	26.2	26.2	25.9	25.8
Montana	Lewis and Clark	29.5	29.4	29.4	29.3	29.2
Montana	Lincoln	35.6	34.8	34.8	33.8	33.7
Montana	Missoula	29.8	28.7	28.6	28.2	28.0
Montana	Sanders	20.1	19.6	19.6	19.3	19.3
Montana	Silver Bow	32.8	32.4	32.4	32.2	32.0
Montana	Yellowstone	18.3	18.3	18.3	18.2	18.1
Nebraska	Douglas	24.3	20.0	19.9	19.3	19.0
Nebraska	Hall	18.3	15.0	14.9	14.5	14.4
Nebraska	Lancaster	18.9	14.4	14.3	14.0	13.9
Nebraska	Sarpy	22.9	17.5	17.5	17.1	16.9
Nebraska	Scotts Bluff	17.6	16.3	16.3	15.9	15.8
Nebraska	Washington	20.8	15.9	15.8	15.3	15.1
Nevada	Clark	23.0	21.4	21.4	20.8	20.5
Nevada	Washoe	34.9	31.5	31.4	30.8	30.4
New Hampshire	Belknap	17.9	11.2	11.2	11.1	11.0
New Hampshire	Cheshire	28.9	26.2	26.2	26.0	26.0
New Hampshire	Grafton	20.5	15.7	15.7	15.6	15.6
New Hampshire	Hillsborough	26.5	22.4	22.4	22.3	22.2
New Hampshire	Merrimack	24.6	20.3	20.3	20.2	20.1
New Hampshire	Rockingham	23.7	18.7	18.7	18.4	18.3
New Hampshire	Sullivan	23.3	19.1	19.1	19.0	19.0
New Jersey	Atlantic	27.4	15.9	15.9	15.8	15.7
New Jersey	Bergen	34.6	19.7	19.5	19.1	18.9
New Jersey	Camden	33.2	22.4	22.3	22.0	21.9
New Jersey	Essex	38.4	24.9	24.9	24.3	24.1
New Jersey	Gloucester	25.7	17.1	17.1	16.8	16.7
New Jersey	Hudson	39.6	27.4	27.3	26.4	26.2
New Jersey	Mercer	32.0	22.0	22.0	21.7	21.6

State	County	2007 Baseline DV	2018 Reference DV	2018 Tier 3 Control DV	2030 Reference DV	2030 Tier 3 Control DV
New Jersey	Middlesex	29.9	18.8	18.8	18.5	18.4
New Jersey	Morris	28.9	17.8	17.7	17.4	17.4
New Jersey	Ocean	28.0	16.9	16.9	16.8	16.7
New Jersey	Passaic	33.3	20.7	20.6	20.2	20.1
New Jersey	Union	37.6	24.3	24.3	23.4	23.4
New Jersey	Warren	33.6	24.3	24.2	23.8	23.7
New Mexico	Bernalillo	16.9	15.3	15.3	15.2	15.0
New Mexico	Chaves	16.2	13.2	13.2	13.8	13.8
New Mexico	Doña Ana	29.4	26.9	26.9	28.1	27.7
New Mexico	Grant	10.1	9.9	9.9	10.5	10.5
New Mexico	Sandoval	15.4	14.4	14.4	14.5	14.4
New Mexico	San Juan	12.5	12.4	12.4	12.4	12.4
New Mexico	Santa Fe	9.1	8.4	8.4	8.5	8.4
New York	Albany	26.5	19.5	19.5	19.1	19.0
New York	Bronx	35.3	25.1	25.0	24.2	24.1
New York	Chautauqua	26.5	13.9	13.9	13.8	13.8
New York	Erie	30.4	20.8	20.7	20.4	20.4
New York	Essex	17.5	9.9	9.8	9.8	9.8
New York	Kings	33.1	21.8	21.8	21.4	21.3
New York	Monroe	27.9	17.8	17.8	17.5	17.4
New York	New York	38.0	27.3	27.3	26.6	26.5
New York	Niagara	28.7	18.4	18.3	18.1	18.0
New York	Onondaga	25.8	15.1	15.1	15.1	15.0
New York	Orange	27.6	17.9	17.8	17.5	17.5
New York	Queens	30.7	21.2	21.1	20.7	20.6
New York	Richmond	31.4	20.1	20.1	19.5	19.5
New York	St. Lawrence	20.3	12.8	12.7	12.7	12.7
New York	Steuben	24.6	14.0	14.0	13.9	13.8
New York	Suffolk	27.4	15.8	15.7	15.6	15.5
New York	Westchester	31.2	18.5	18.4	18.1	18.0
North Carolina	Alamance	28.5	18.6	18.6	18.2	18.0
North Carolina	Buncombe	26.7	15.1	15.0	14.8	14.7
North Carolina	Caswell	26.6	16.4	16.3	15.9	15.8
North Carolina	Catawba	29.5	17.4	17.4	17.1	16.9
North Carolina	Chatham	25.0	15.5	15.4	15.2	15.1
North Carolina	Cumberland	27.5	17.7	17.7	17.4	17.3
North Carolina	Davidson	28.5	16.6	16.5	16.3	16.1
North Carolina	Duplin	24.1	13.4	13.4	13.4	13.4

State	County	2007 Baseline DV	2018 Reference DV	2018 Tier 3 Control DV	2030 Reference DV	2030 Tier 3 Control DV
North Carolina	Durham	30.0	17.4	17.4	17.0	16.8
North Carolina	Edgecombe	24.6	16.0	16.0	15.8	15.7
North Carolina	Forsyth	28.4	17.5	17.5	17.2	17.0
North Carolina	Gaston	27.5	15.4	15.4	15.2	15.1
North Carolina	Guilford	24.1	15.9	15.9	15.5	15.4
North Carolina	Haywood	28.5	19.3	19.2	19.1	19.0
North Carolina	Lenoir	23.0	13.3	13.3	13.3	13.2
North Carolina	McDowell	28.0	16.9	16.8	16.6	16.6
North Carolina	Martin	22.1	15.2	15.2	15.2	15.2
North Carolina	Mecklenburg	28.9	17.2	17.2	16.9	16.8
North Carolina	Mitchell	27.3	17.9	17.8	17.7	17.6
North Carolina	Montgomery	25.4	14.5	14.5	14.3	14.3
North Carolina	New Hanover	25.4	15.3	15.2	15.2	15.2
North Carolina	Onslow	24.7	14.6	14.6	14.5	14.5
North Carolina	Orange	29.0	16.9	16.9	16.4	16.3
North Carolina	Pitt	24.7	15.9	15.9	15.8	15.7
North Carolina	Robeson	26.8	18.2	18.2	18.0	17.9
North Carolina	Rowan	27.5	17.0	16.9	16.7	16.6
North Carolina	Swain	26.0	16.5	16.5	16.3	16.3
North Carolina	Wake	29.1	18.0	17.9	17.7	17.5
North Carolina	Watauga	25.2	13.7	13.7	13.6	13.5
North Carolina	Wayne	27.2	16.0	15.9	15.8	15.8
North Dakota	Billings	12.8	11.8	11.8	11.7	11.6
North Dakota	Burleigh	16.1	14.6	14.6	14.3	14.3
North Dakota	Cass	19.1	16.2	16.2	15.6	15.5
North Dakota	Mercer	15.1	13.5	13.5	13.2	13.2
Ohio	Athens	30.8	15.6	15.6	15.4	15.4
Ohio	Butler	33.3	22.3	22.3	22.0	21.8
Ohio	Clark	33.1	22.8	22.8	22.4	22.2
Ohio	Clermont	30.1	17.1	17.1	16.9	16.7
Ohio	Cuyahoga	39.0	28.2	28.1	28.2	28.0
Ohio	Franklin	33.3	22.2	22.1	21.8	21.6
Ohio	Greene	29.9	18.9	18.9	18.6	18.4
Ohio	Hamilton	34.6	24.5	24.4	23.9	23.6
Ohio	Jefferson	37.0	22.7	22.7	22.5	22.4
Ohio	Lake	31.7	17.5	17.5	17.4	17.3
Ohio	Lawrence	34.8	20.9	20.9	20.6	20.5
Ohio	Lorain	30.7	20.9	20.9	20.7	20.6

State	County	2007 Baseline DV	2018 Reference DV	2018 Tier 3 Control DV	2030 Reference DV	2030 Tier 3 Control DV
Ohio	Lucas	34.7	24.9	24.8	24.3	24.0
Ohio	Mahoning	32.8	21.6	21.5	21.3	21.0
Ohio	Medina	28.7	19.6	19.6	19.4	19.2
Ohio	Montgomery	33.7	21.0	20.9	20.7	20.5
Ohio	Portage	30.9	19.5	19.5	19.3	19.2
Ohio	Preble	29.8	20.8	20.8	20.5	20.3
Ohio	Scioto	31.6	19.9	19.9	19.7	19.6
Ohio	Stark	36.0	23.3	23.2	23.0	22.8
Ohio	Summit	34.7	23.6	23.5	23.1	22.9
Ohio	Trumbull	33.2	20.9	20.9	20.6	20.4
Ohio	Warren	27.1	18.4	18.4	18.2	18.0
Oklahoma	Caddo	26.2	20.3	20.3	20.1	19.9
Oklahoma	Cherokee	27.4	20.5	20.4	20.3	20.2
Oklahoma	Kay	26.8	21.9	21.8	21.6	21.5
Oklahoma	Mayes	25.4	18.1	18.1	17.8	17.8
Oklahoma	Muskogee	27.5	20.8	20.8	20.7	20.6
Oklahoma	Oklahoma	24.2	18.7	18.7	18.7	18.5
Oklahoma	Ottawa	24.9	18.1	18.1	17.9	17.7
Oklahoma	Pittsburg	24.8	18.4	18.4	18.4	18.3
Oklahoma	Sequoyah	27.3	22.0	22.0	22.0	21.9
Oklahoma	Tulsa	27.4	20.8	20.8	20.7	20.5
Oregon	Harney	33.0	35.6	35.6	35.0	34.9
Oregon	Jackson	33.2	33.2	33.1	32.1	32.0
Oregon	Josephine	30.6	31.8	31.7	31.0	31.0
Oregon	Klamath	46.1	46.6	46.5	45.4	45.3
Oregon	Lake	41.4	42.8	42.8	42.2	42.2
Oregon	Lane	42.4	41.1	41.0	39.9	39.8
Oregon	Multnomah	29.1	28.5	28.5	27.4	27.4
Oregon	Umatilla	24.7	24.7	24.7	23.6	23.5
Oregon	Union	21.7	21.2	21.2	20.4	20.4
Oregon	Washington	31.6	32.7	32.8	31.9	31.9
Pennsylvania	Adams	31.4	19.3	19.2	19.0	18.9
Pennsylvania	Allegheny	54.4	40.1	40.1	39.5	39.5
Pennsylvania	Beaver	37.0	25.0	25.0	24.8	24.8
Pennsylvania	Berks	34.1	27.2	27.1	26.6	26.4
Pennsylvania	Bucks	32.9	23.1	23.1	22.8	22.7
Pennsylvania	Cambria	35.3	19.5	19.4	19.2	19.2
Pennsylvania	Centre	31.6	19.0	19.0	18.9	18.8

State	County	2007 Baseline DV	2018 Reference DV	2018 Tier 3 Control DV	2030 Reference DV	2030 Tier 3 Control DV
Pennsylvania	Chester	36.4	24.1	24.1	23.7	23.6
Pennsylvania	Cumberland	34.4	24.9	24.9	24.2	24.1
Pennsylvania	Dauphin	35.8	26.5	26.4	25.6	25.4
Pennsylvania	Delaware	33.0	22.6	22.5	22.0	22.0
Pennsylvania	Erie	30.9	19.2	19.2	19.1	19.0
Pennsylvania	Lackawanna	29.4	19.9	19.8	19.5	19.4
Pennsylvania	Lancaster	37.0	31.1	31.1	30.6	30.3
Pennsylvania	Mercer	29.8	19.1	19.1	18.8	18.7
Pennsylvania	Montgomery	28.5	19.9	19.8	19.6	19.5
Pennsylvania	Northampton	35.8	26.3	26.2	25.8	25.7
Pennsylvania	Philadelphia	36.6	25.0	25.0	24.5	24.4
Pennsylvania	Washington	33.9	19.3	19.3	19.0	18.9
Pennsylvania	Westmoreland	35.2	20.1	20.1	19.9	19.8
Pennsylvania	York	34.6	27.0	27.0	26.5	26.4
Rhode Island	Kent	23.1	13.7	13.7	13.6	13.6
Rhode Island	Providence	28.2	21.8	21.8	21.6	21.6
South Carolina	Charleston	23.1	15.7	15.7	15.7	15.7
South Carolina	Chesterfield	24.9	16.2	16.2	16.0	15.9
South Carolina	Edgefield	26.8	18.0	18.0	17.8	17.8
South Carolina	Florence	26.7	17.1	17.0	16.8	16.7
South Carolina	Greenville	30.4	21.3	21.3	20.9	20.7
South Carolina	Greenwood	29.3	18.4	18.3	18.1	18.0
South Carolina	Horry	29.2	21.0	21.0	20.9	20.8
South Carolina	Lexington	28.4	18.6	18.5	18.1	18.0
South Carolina	Oconee	23.3	13.3	13.2	13.1	13.0
South Carolina	Richland	28.5	18.4	18.3	18.1	18.0
South Carolina	Spartanburg	28.5	17.3	17.3	17.0	16.9
South Dakota	Brookings	21.6	16.7	16.6	16.1	16.0
South Dakota	Brown	17.5	14.6	14.5	14.1	14.0
South Dakota	Codington	23.9	18.6	18.5	17.8	17.7
South Dakota	Custer	14.1	13.2	13.2	13.2	13.2
South Dakota	Jackson	12.4	11.6	11.6	11.6	11.6
South Dakota	Minnehaha	25.5	19.8	19.6	18.5	18.2
South Dakota	Pennington	17.4	16.4	16.4	16.4	16.3
Tennessee	Blount	31.0	20.3	20.2	20.0	19.8
Tennessee	Davidson	31.5	21.3	21.3	20.9	20.7
Tennessee	Dyer	28.9	16.7	16.7	16.4	16.3
Tennessee	Hamilton	31.3	19.6	19.6	19.1	19.0

State	County	2007 Baseline DV	2018 Reference DV	2018 Tier 3 Control DV	2030 Reference DV	2030 Tier 3 Control DV
Tennessee	Knox	32.6	20.0	20.0	19.5	19.2
Tennessee	Lawrence	29.6	18.1	18.0	17.9	17.8
Tennessee	Loudon	31.0	20.3	20.3	20.0	19.9
Tennessee	McMinn	32.8	19.7	19.7	19.4	19.3
Tennessee	Madison	28.1	16.4	16.3	16.2	16.1
Tennessee	Maury	28.1	16.5	16.5	16.2	16.1
Tennessee	Montgomery	32.6	18.4	18.3	18.2	18.1
Tennessee	Putnam	25.5	15.5	15.5	15.3	15.2
Tennessee	Roane	29.0	17.6	17.6	17.2	17.1
Tennessee	Shelby	33.5	19.4	19.4	19.0	18.9
Tennessee	Sullivan	29.4	17.4	17.4	17.1	17.0
Tennessee	Sumner	29.6	19.4	19.4	19.0	18.8
Texas	Bowie	27.2	19.1	19.1	18.9	18.8
Texas	Dallas	23.6	16.9	16.8	16.8	16.6
Texas	Ector	17.4	15.6	15.5	16.2	16.1
Texas	El Paso	27.1	26.9	26.9	30.8	30.7
Texas	Harris	29.8	22.4	22.3	22.2	22.1
Texas	Harrison	23.4	17.4	17.4	17.0	17.0
Texas	Hidalgo	24.3	21.4	21.4	21.6	21.6
Texas	Nueces	27.8	21.0	21.0	21.1	21.0
Texas	Orange	28.7	22.4	22.3	22.4	22.4
Texas	Potter	14.8	13.2	13.2	13.5	13.4
Texas	Tarrant	24.5	17.7	17.7	17.8	17.5
Texas	Travis	20.9	15.4	15.3	15.2	15.1
Utah	Box Elder	33.8	29.7	29.5	28.2	27.6
Utah	Cache	39.3	34.0	33.9	32.3	31.6
Utah	Davis	37.1	32.8	32.7	31.5	30.9
Utah	Salt Lake	47.5	43.0	42.8	41.5	40.7
Utah	Tooele	25.1	22.0	21.9	21.0	20.7
Utah	Utah	46.1	41.0	40.7	39.0	38.4
Utah	Weber	37.5	32.9	32.7	31.1	30.5
Vermont	Bennington	23.2	14.6	14.6	14.4	14.4
Vermont	Chittenden	25.9	20.4	20.4	20.3	20.2
Vermont	Rutland	28.9	29.2	29.2	29.0	28.9
Virginia	Arlington	29.6	17.8	17.8	17.7	17.5
Virginia	Charles City	28.1	15.4	15.4	15.2	15.1
Virginia	Chesterfield	27.7	15.4	15.3	15.2	15.1
Virginia	Fairfax	31.1	17.8	17.8	17.9	17.7

State	County	2007 Baseline DV	2018 Reference DV	2018 Tier 3 Control DV	2030 Reference DV	2030 Tier 3 Control DV
Virginia	Henrico	29.0	15.7	15.6	15.5	15.4
Virginia	Loudoun	29.0	15.7	15.6	15.6	15.5
Virginia	Page	27.7	14.6	14.5	14.4	14.4
Virginia	Rockingham	26.1	15.8	15.8	15.5	15.4
Virginia	Bristol city	27.5	16.6	16.5	16.4	16.3
Virginia	Hampton city	29.0	16.5	16.5	16.4	16.3
Virginia	Lynchburg city	27.9	16.0	16.0	15.8	15.7
Virginia	Norfolk city	28.0	18.6	18.6	18.3	18.2
Virginia	Roanoke city	31.0	18.2	18.2	17.9	17.7
Virginia	Virginia Beach City	31.1	19.4	19.4	19.3	19.2
Washington	King	31.0	31.0	31.1	30.5	30.6
Washington	Pierce	44.2	44.1	44.1	43.2	43.3
Washington	Snohomish	34.2	33.7	33.7	33.2	33.2
Washington	Spokane	30.1	29.3	29.3	28.1	28.0
Washington	Yakima	37.2	35.8	35.7	32.5	32.4
West Virginia	Berkeley	31.2	22.1	22.1	21.9	21.7
West Virginia	Brooke	40.4	23.9	23.8	23.6	23.5
West Virginia	Cabell	32.9	18.1	18.1	17.9	17.8
West Virginia	Hancock	38.0	19.8	19.8	19.6	19.5
West Virginia	Harrison	30.2	14.2	14.2	14.1	14.0
West Virginia	Kanawha	35.2	17.3	17.3	17.1	17.0
West Virginia	Marion	31.1	16.0	15.9	15.8	15.7
West Virginia	Marshall	33.2	19.0	19.0	19.0	18.9
West Virginia	Monongalia	33.3	13.3	13.3	13.3	13.2
West Virginia	Ohio	30.6	17.0	17.0	16.9	16.8
West Virginia	Raleigh	27.0	13.4	13.4	13.3	13.3
West Virginia	Wood	33.9	19.0	19.0	18.9	18.8
Wisconsin	Ashland	19.0	14.6	14.6	14.5	14.4
Wisconsin	Brown	35.4	31.3	31.1	32.0	31.6
Wisconsin	Dane	34.7	30.7	30.6	30.1	29.8
Wisconsin	Dodge	28.7	25.8	25.7	25.5	25.3
Wisconsin	Forest	20.9	16.3	16.2	16.0	15.9
Wisconsin	Grant	34.5	31.4	31.2	30.4	30.0
Wisconsin	Kenosha	32.1	26.4	26.3	25.7	25.3
Wisconsin	La Crosse	32.1	30.0	29.9	29.2	28.8
Wisconsin	Manitowoc	29.6	25.4	25.3	25.2	25.0
Wisconsin	Milwaukee	37.2	32.2	32.1	31.9	31.4

State	County	2007 Baseline DV	2018 Reference DV	2018 Tier 3 Control DV	2030 Reference DV	2030 Tier 3 Control DV
Wisconsin	Outagamie	32.8	28.4	28.2	27.9	27.5
Wisconsin	Ozaukee	31.7	27.3	27.2	26.6	26.3
Wisconsin	St. Croix	26.7	23.5	23.5	22.8	22.5
Wisconsin	Sauk	28.1	25.1	25.0	24.5	24.3
Wisconsin	Taylor	27.7	22.7	22.6	22.0	21.7
Wisconsin	Vilas	26.5	21.5	21.4	20.9	20.7
Wisconsin	Waukesha	32.3	27.7	27.6	27.1	26.7
Wyoming	Campbell	14.0	13.5	13.5	13.3	13.3
Wyoming	Converse	9.8	9.2	9.2	9.1	9.1
Wyoming	Fremont	26.2	25.2	25.2	25.0	24.9
Wyoming	Laramie	10.2	9.1	9.1	9.0	9.0
Wyoming	Sheridan	25.7	25.0	25.1	24.8	24.7

United States
Environmental Protection
Agency

Office of Air Quality Planning and Standards
Air Quality Assessment Division
Research Triangle Park, NC

Publication No. EPA-454/R-14-002
February, 2014

www.ingramcontent.com/pod-product-compliance
Lightning Source LLC
Chambersburg PA
CBHW081457170526
45166CB00008B/2463